THE COSMIC–QUANTUM
CONUNDRUM

ALSO BY PETER STRASSBERG, M.D.

The One-Hundred Year Diet
The Hunger for More: God, Gravity, and the Big Bang

THE COSMIC–QUANTUM CONUNDRUM

If you can explain it,
You don't understand it.

Peter Strassberg, M.D.

Full Court Press
Englewood Cliffs, New Jersey

Published in the United States of America
by Full Court Press, 601 Palisade Avenue,
Englewood Cliffs, NJ 07632
fullcourtpressnj.com

ISBN 978-1-938812-79-8
Library of Congress Catalog No. 2016948445

*Book design by Barry Sheinkopf for Bookshapers
(bookshapers.com)*

Cover art courtesy istockphoto.com

Colophon by Liz Sedlack

TO PHIL
in memoriam

PREFACE

This work is both a companion to, and a clarification of, many of the ideas initially put forth in my prior book— *The Hunger for More: God, Gravity, and the Big Bang.* It attempts a slightly more formal explanation using the concept of surfaces and interiors, or spheres and balls. As before, the math employed is relatively simple and, in order to not bother the reader, generally relegated to the appendices.

The basic concept is the same. The Big Bang has become a clumsy contrivance, an explanation held together with ad hoc, almost surreal appendages of impossible inflation and chimerical dark energy. A more direct, and simpler idea is that of a 3-sphere, or three-dimensional exterior on a higher-dimensional orb. The mathematics of such a surface utilizes the straightforward relationship of a circle to its tangent. It is easily understood and should, therefore, allow for the unfolding of a more rational explanation.

However, once we do away with an initial, cataclysmic event, we also have to revise the Standard Model of particle formation. Current belief ascribes the entire panoply of the quantum world to growth and evolution as the universe expands and cools. But, since there was no Big Bang, no primary occurrence, there is no inherent basis for this explanation. Particles are, instead, shown to be the surface "nodes" of the immense 3-sphere we call the universe.

Finally, the weirdness of the quantum world is brought, somewhat, into focus. It becomes, hopefully, understandable by allowing spin (an imaginary motion that underlies most of quanta's strange actions) to be a movement into and out of the higher dimension. By so defining this ever present, constant rotation (through what, to us, appears unreal) we can attempt an explanation of the multiple bizarre findings in the submicroscopic world.

The book, then, is a clear break with what are today's accepted beliefs. It opens us to a unique view of both the very large and the vanishingly small. Hopefully it has uncovered some kernel of truth that will subsequently flourish, in the hands of more accomplished individuals, as a branch on the forever evolving tree of science.

—*P.S.*

TABLE OF CONTENTS

Part One: Cosmic Concerns

Part Two: Quantum Queries

Part Three: Final Formulations

PART ONE

Cosmic Concerns

Chapter 1

THE BIG BANG

Its history and problems

AN INITIAL EVENT

I T IS ALWAYS FOOLISH TO SEEK what is unattainable, and the title of this book exposes this inherent conceit. However, much of modern scientific theory has gone so far off a logical track that one must still attempt this quixotic journey. So with a large helping of humility and just a touch of temerity, we begin.

A significant amount of today's science is based on the Big Bang theory—the belief in a cataclysmic start to our world. This concept goes back almost a hundred years to a Jesuit priest—Georges Lemaître—who was also a professor of physics. In the 1920s he first conceived the idea—an initial event, an exploding "atom," that contained the seeds of what has become our universe.

His views were not immediately acknowledged; however, over a period of several decades they eventually coalesced into current theory. The real reason for their acceptance was the work performed by Edwin Hubble, a leading astronomer of the time, who, by using the 100-inch Mount Wilson telescope, could see farther and clearer than anyone else. He was able to conclusively prove that our Milky Way was but one galaxy in a giant cosmos of many. More significantly, he also noted that the light coming from ever-distant galaxies continuously shifted toward longer wavelengths—it "redshifted."

DOPPLER EFFECT

Earlier astronomers, working before Hubble, had already concluded that a shift in the wavelength of light was a Doppler effect. Wavelengths, from any source, moving toward a person would shorten; those moving away would lengthen. We see this in the wail of an ambulance's siren that as it approaches is higher pitched (greater frequency, shorter wavelength) but lower as it departs (lesser frequency, longer wavelength).

Prior astronomers had noted that stars in our own galaxy moving toward us shifted to the blue, or shorter, wavelengths, and that those moving away toward the red, or longer. When Hubble measured these changes in distant galaxies, they invariably shifted *only* toward the red spectrum; and what was even more interesting, the farther away, the greater was this distortion. Since the Doppler effect had long been understood to not only reveal the direction of flight but also the speed of the object (the greater the shift, the faster the motion), Hubble's results were interpreted as showing a direct link between distance and velocity (all objects were moving away from all others, the farther away the faster).

This information was then used to establish the scientific basis for Lemaître's beliefs. The concept is that, since everything is moving away from everything else—ever-farther, ever-faster—if one were simply to run time backwards, all things would be ever-closer until, finally, all would be congealed into one spot, one giant atom.

STEADY STATE THEORY

The difficulty with Hubble's initial findings, however, was that the universe could only be shown to be about two billion years old. Scientists had already estimated the age of the Earth at close to five billion years, so this obvious discrepancy caused a significant problem for a Big Bang interpretation. Therefore, for almost forty years, the theory languished as other equally plausible scenarios were put forth. A major early competitor was the Steady State theory, conceived by Fred Hoyle, an influential physicist and cosmologist. In that concept, the universe had no beginning but had always been as it currently appeared. Expansion was forever ongoing, with a steady, albeit very slight, continuous formation of new hydrogen atoms. As in most debates, the less real information available, the more contentious were the arguments.

COSMIC MICROWAVE BACKGROUND (CMB)

This back-and-forth between major theories continued through the mid 1960s until two scientists at Bell Labs, Arno Penzias and Robert Wilson, while trying to eliminate some static or noise in their radio telescope, made a major, serendipitous discovery. To uncover the source of this annoying interference, they had gone so far as to eradicate the pigeons roosting in the telescope's massive structure, even cleaning their droppings; however, the problem would not resolve. They concluded, finally, that this static or noise was *real*—that there existed, throughout the universe, an underlying microwave radiation equal in intensity in all directions.

Big Bang adherents seized upon this information as proof of their theory. They had, some time before, estimated that, if the Big Bang had taken place, infrared radiation, photons approximately 10^{-6} meters in length, would have spewed out from the initial event. Over the resulting billions of years, as the universe had expanded many times, these photons would have lengthened about 1000-fold, going from around 10^{-6} m to 10^{-3} m, or from infrared to microwave size. Hence, the cosmic microwave background (CMB) now found was explained as the remnant of the initial "explosion." Since the Steady State proponents had no real alternative for this finding, their theory became less attractive and was, essentially, abandoned.

INFLATION

The Big Bang continued as the predominant theory for the next ten to fifteen years; however, in the late 1970s other disquieting problems arose. Supporters could not explain why the universe was as homogeneous as it appeared, for, if there had *been* a big, "messy" initial event, the current world would appear less uniform than it does. Also, *relic particles*, hypothetical entities supposedly formed in a very early universe, should still be present but could not be found. Finally, CMB was uniform throughout, in any direction one looked; however, if there had been a Big Bang, there would not have been sufficient time for all parts to have interacted or comingled. So why should CMB be equal throughout?

A theoretical physicist, Alan Guth, then came up with an intriguing concept that appeared to account for all these inconsistencies. He was writing in the late 1970s, shortly after the end of America's Vietnamese engagement, and our country was paying off its war debts by the time-honored method of inflating its money supply. He postulated that, in a similar fashion, the universe, very soon after the Big Bang event, inflated (to an enormous size in a very short period of time). He thereby solved all the problems associated with uniformity and lack of particles; for just as a crinkly balloon, once blown up, becomes smooth, so too did the universe. No matter what it may have resembled following the cataclysmic start, it evened out in a very rapid and immense inflationary process.

Therefore, CMB should be the same throughout, as earliest contact, prior to inflation, would be maintained. In a similar manner, relic particles—*magnetic monopoles*—although sufficient in initial quantity to satisfy most theories, were now so diffusely spread out as to essentially have disappeared. Although Guth could not give a real or coherent reason for this sudden inflation, if it *had* occurred it would answer the many questions that had been raised. The concept caught on, perhaps because it had been proposed in an era of fiscal impropriety; but for whatever reason, it became accepted gospel.

DARK ENERGY

Thus, into the 1980s and '90s, the Big Bang continued as the major theory. In the late 1990s, another question arose. If there had been a Big Bang causing expansion, was that expansion perhaps now slowing or, possibly, even contracting? One should remember, gravity was the only force known to function on a cosmic scale, and it was considered accumulative, not expansive. Therefore, had the initial cataclysmic energy perhaps sufficiently diminished, and was gravity now causing a reversal?

In order to explore this idea, scientists sought a form of "distance marker" or *standard candle* to determine true celestial expanse. It is very difficult to really judge how far away an object is in space. Up to then, astronomers had employed the redshift as a crude metric, or crutch, for measurement. According to Hubble's law, the universe expands at a steady rate—a little less than 70 kilometers/second for each additional 3.26 million light-years of distance. So by measuring the redshift of an entity and thereby determining its velocity, its distance could be estimated. However, if one had a standard candle, in this case an extremely luminous supernova (type 1a), which was visible for many billions of light-years, and which always shone with the same intensity, one could determine, by how dim it had become, how far away it really was. Then, by evaluating its redshift, one could set up a more precise tabulation with distance, that is, a more accurate version than Hubble's constant.

However, when these supernovas were discovered, to the bewilderment of the astronomers, they were actually *farther away* than Hubble's theory allowed. Hence the universe was not only still expanding, it was doing so ever faster. Therefore the cosmos had apparently been growing at a more rapid rate for the last 6 to 10 billion years than originally assumed; and a new source of energy to fuel this change was required. Since no known source was evident, this new entity was deemed "unknown" or (perhaps not as politically correct, but catchier) *dark energy*. Its cause was not understood; but, inasmuch as Einstein had previously used, then discarded, a fudge factor—the *cosmological constant*—to make the universe stable, this afterthought was resurrected and employed by many as its basis.

BASIC PROBLEM

Today we therefore have a theory of the universe—the Big Bang—founded on an idea put forth close to a hundred years ago, by a priest-scientist, Georges Lemaître, and supported by the redshift discoveries of Edwin Hubble. The theory is based on the obvious Doppler effect seen in nearby stars with red and blue changes, and then supposedly observed in distant galaxies with ever-increasing redshifts. It was strongly buttressed in the 1965 finding of CMB, by Penzias and Wilson, and then later supplemented by Alan Guth's theoretical concept of inflation. Finally, it has been further tweaked with increased expansion caused by an entirely new form of energy, unknown yet believed to exist.

The theory is reasonable, even if it requires the *ad hoc* addition of inflation and a new, hitherto unsuspected form of energy. It is still the best fit for Hubble's initial findings, *if one assumes that the redshift is due to a Doppler effect*— hence signifying expansion.

But what if the very foundation of this construct is incorrect? *What if the assumption of a Doppler effect is wrong*? What then happens to this theory?

If the redshift is *not* a Doppler effect, then the concept requires a complete revision. But into what?

The following several chapters will hopefully reveal an alternative scenario, and in so doing bring a fitting finality to what is becoming a tottering and evermore precarious edifice—the Big Bang theory.

Chapter 2

SURFACES

Our world as the enclosure
to a higher dimension

SPHERES AND BALLS

AS PREVIOUSLY STATED, the Big Bang theory is the most appropriate structure that allows for an understanding of the universe. At its core is Hubble's seminal work showing a redshift ever-increasing with distance.

This finding was attributed by Hubble and others to a Doppler effect; hence, the greater the distance and redshift, the greater the velocity. The universe, by such a measure, has constantly expanded since its inception, some 13.8 billion years ago. In fact, given dark energy, its expansion has actually increased over the last 6 to 10 billion or so years.

I wish, nevertheless, to assign an entirely different meaning to the increased redshift. Before doing so, however, a few basic mathematical definitions are in order.

In mathematics, if one formally wants to describe what to all of us is merely a "circle," its circumference becomes a "1-sphere" and the area contained within a "2-ball." Thus, in math, surfaces of round objects are called *spheres*, the objects they enclose *balls*. Let us give a few examples.

In zero dimensions, the entire entity is simply a point, and its mathematical formula is denoted as 1. It has no cover, as it has nothing to enclose.

In one-dimensional space, the object enclosed becomes a straight line. It has a point at both ends; this is its cover. Since each point is denoted as 1, a point at either end would equal 2. The line enclosed is divided in half, and each part is considered its radius, or r. Hence, the formula for the one-dimensional ball (1-ball) is 2r, and for the zero-dimensional cover (0-sphere) is 2. Remember, the surface is always one dimension less than what is covered:

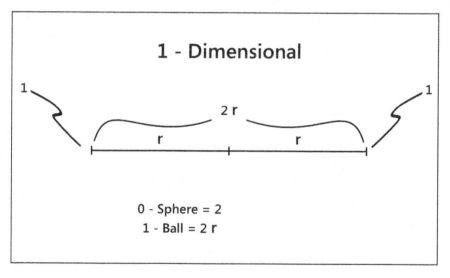

In two-dimensional space, as we have already noted, a circle has a circumference (1-sphere) and an enclosed area (2-ball). The formula for the circumference is $2\pi r$ and for the area, πr^2. Pi (π) is, simply put, a ratio of the circumference of any circle to its diameter. Inasmuch as the circumference cannot be absolutely defined as a number (a straight line can be accurately measured, whereas a curved line cannot), the *ratio* of a non-measurable or indefinite number and a real number is considered to be *irrational*. Hence π has an infinite and random sequence; it continues forever, never duplicating itself (we typically use 3.14159 as an approximation).

The only reason to discuss π to such a degree is that, whenever it appears in an equation, a circle of some kind is envisaged. We will see that, as one goes into ever-higher dimensions, spheres follow a simple formula: they are always the product of a circle (our 1-sphere—$2\pi r$) multiplied by the "interior" of a round object, two dimensions smaller than that enclosed. Let us show that two-dimensional subject:

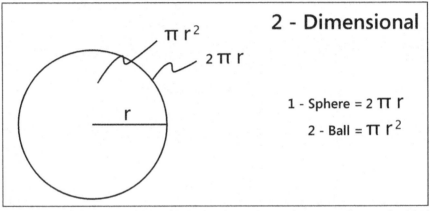

If we now go to normal three-dimensional space, the surface of what we would commonly call a globe is, mathematically speaking, a 2-sphere. It has one dimension less than what it covers, a 3-ball. The formula for a 2-sphere is $4\pi r^2$, as in the example on the next page.

Finally, let us go one dimension further, to fourth-dimensional space. In mathematics, one can go to any number of dimensions, whether or not they exist in the real world. The formula for spheres holds no matter how high we go; however, we will stop at four. The surface of a fourth-dimensional round object, or 4-ball, is a 3-sphere (again, the surface, by definition, is always one dimension less). Its formula is $2\pi^2 r^3$.

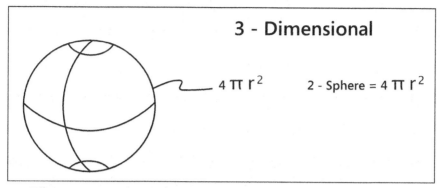

This is as much math as will be used in this book; for I know every time an equation is written, half of my readers depart. The remainder of the math thankfully appears in the Appendices, to be read only by those who are so inclined or have difficulty sleeping. (Please see Appendix A.)

ONE-DIMENSIONAL WORLD

Now, if we lived in a one-dimensional world, if we were just segments of a straight line, how would we visualize a higher dimension? Let me picture our one-dimensional universe:

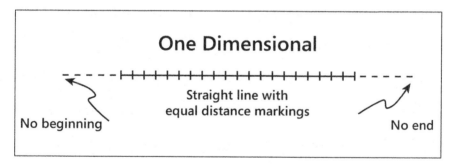

But what if this one-dimensional world really was on a *curved* line and thus *two*-dimensional? How would inhabitants of what is considered a straight line envisage their world if it really were round? If we draw it, we get the image at the top of the next page.

Each distance on the curved line is projected as longer on the

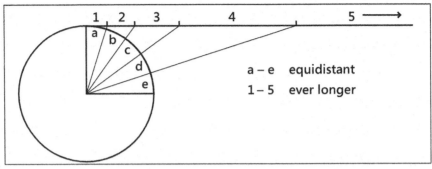

straight line. And the farther we travel in higher two-dimensional space, the greater these distortions appear in one-dimensional space.

MEASURING WITH LIGHT WAVES

If we were to use light waves as measuring rods (and any consistent metric can be a yardstick), since the same distance is visualized as continuously elongating (as we travel farther on the curved line) the distance marker must compensate by stretching. Although that marker is always recording the same interval on the two-dimensional surface, it is appreciated as ever-lengthening on the one-dimensional line; therefore, the light waves, to maintain the same total number, must get ever-larger. So, if we redraw our ruler as these waves, we get the following altered image:

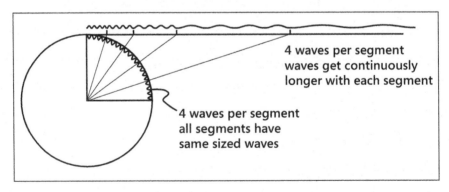

Each time the same distance is drawn as four individual light waves; however, one sees that, to fill the apparent increase in distance, those waves *have to* elongate—they redshift. The waves, although exactly the same size, appear ever-longer as we journey farther out on the curve.

THREE-DIMENSIONAL WORLD

Up to now we have drawn a one-dimensional (straight-line) universe that really lies on a two-dimensional curve, or circle. Let us extrapolate that to our *three*-dimensional world. To get perspective we start with a cube:

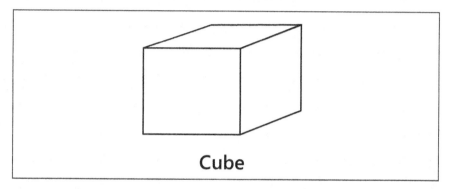

Cube

We then partially flatten it front to back so that it is more two-dimensional in appearance; we now have a tablet:

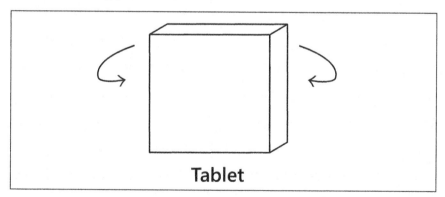

Tablet

We then shorten it top to bottom so that it looks almost one-dimensional; we now have a rod:

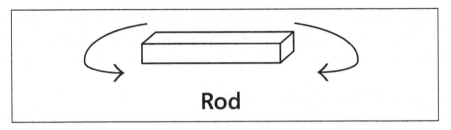

Rod

We still have a three-dimensional object; it just appears rather one-dimensional. We then draw it as a straight line. Thus, we have the following:

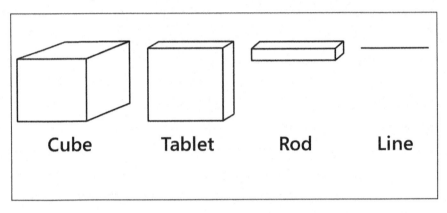

Cube **Tablet** **Rod** **Line**

I am implying that, *for drawing purposes only*, the straight line is really a three-dimensional cube. Therefore, I am stating, again simply for visualization, that what before was a one-dimensional world really represents our three-dimensional universe. If this perception is acceptable, then the circle, one dimension greater than the line, would still remain as such (one dimension higher), and what we spoke of before as the elongation of light waves in a one-dimensional world (residing on the surface of a two-dimensional object) would now become a similar lengthening, but in a *three*-dimensional realm (residing on a *four*-dimensional object): a 3-sphere on a 4-ball. So, the illustration allows us to conceive of a straight line as a full three-dimensional domain.

VISUALIZING A 3-SPHERE

Another way to envisage this would again be to distort our three-dimensional world and place it on a sphere. Once more we begin with a cube of space:

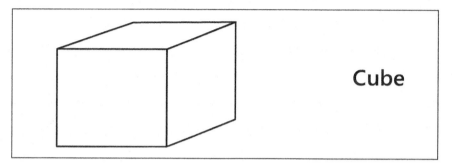

Cube

We then flatten it into a tablet:

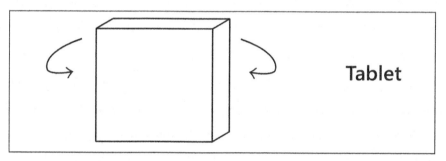

Tablet

It remains perfectly three-dimensional, only thinner. If we then bend it, what do we get?

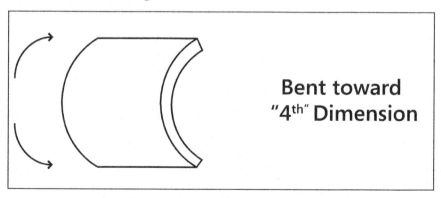

Bent toward "4th" Dimension

We have taken a flat three-dimensional tablet and bent it into a higher dimension. Of course, this is impossible to truly visualize; but by using the above trick, we can at least *imagine* it. If we were to add equal-distance markers, of any kind, similar to those in our previous example of light waves, what happens?

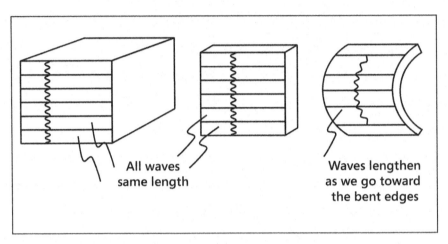

All waves same length

Waves lengthen as we go toward the bent edges

We see a widening of these markers as we go to the periphery of our bent three-dimensional tablet; and if they were light waves, they would have redshifted. If we now superimpose this bent shape upon a sphere, we are making an initial attempt at drawing a 3-sphere:

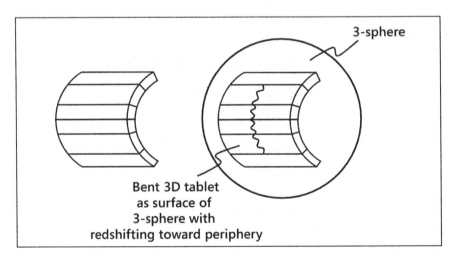

3-sphere

Bent 3D tablet as surface of 3-sphere with redshifting toward periphery

We are thus illustrating a 3-sphere; and its three-dimensional surface would continuously redshift the farther away it was viewed from any point.

Hence, what I am saying is that our world is a 3-sphere, the three-dimensional surface of a 4-ball (an unknowable fourth-dimensional object, a black hole). The farther out we look, in any direction, the more redshifted our world appears. I'm arguing that, when Hubble evaluated distant galaxies and computed their redshifts, he was *not* measuring the Doppler effect seen in nearby stars; instead, as he was looking ever so far, he was beginning to view *the actual curve of the universe*. Astronomers can now peer out over 13 billion light-years; thus the redshift becomes much more pronounced. What was a fractional change to Hubble is, today, a change in magnitude of 10, or more, times.

OTHER DIMENSIONAL DISTORTIONS

Finally, let me demonstrate how this increase in size holds in any dimension. If we were to go from a three-dimensional world to a flat plane, we again see the distortion:

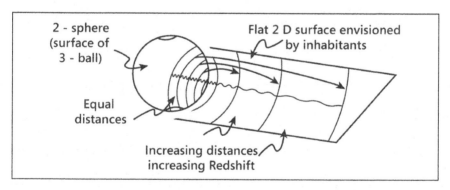

Ever-widened circles that are the same distance apart on the 2-sphere are increasingly redshifted when projected onto the flat plane. The classical map of the world most of us saw in school—

the Mercator projection—greatly distorts all land masses north and south of the equator in precisely the same way:

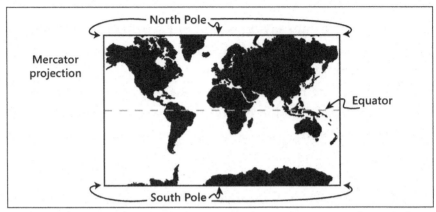

Greenland becomes much larger in this projection of a 2-sphere (three-dimensional cover) onto a flat surface, and the North and South Poles cover the entire width of the map. Let me illustrate:

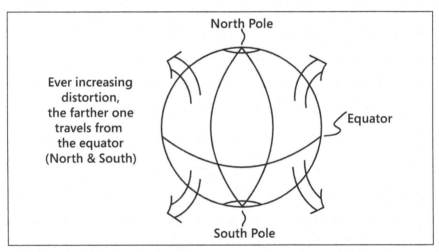

So the redshift increase explained by Hubble as a Doppler effect, when seen in ever-farther galaxies, is in reality only the effect of a higher-dimensional curve.

Chapter 3

DEBUNKING THE "THEORY"

*The demise of dark energy and inflation,
and an alternative explanation
for the cosmic microwave background*

Z PARAMETER

I F WE TAKE THE REDSHIFT TO MEAN a curve into fourth-dimensional space, then our world becomes the surface of that fourth-dimensional object—the 3-sphere covering a 4-ball of vast, unknowable magnitude and shape.

When astronomers discuss redshift changes, they use the concept of the z parameter. It is, simply put, the amount the redshift increases with distance. It tells us how many times that distant object's light has shifted to a longer wave.

To find z, you take the new redshift, subtract the original redshift, and divide what is left by the original redshift. For example, if a light wave from deep space is 1000 nanometers (1000 x 10^{-9} meters) but on Earth 500 nm, then the z factor is:

1000 - 500/500 = 1, or a 100 percent increase.

Thus, it is two times larger than the initial wave.

In a like manner, a z factor of 2.0 is three times as great, and a z of 0.1 is 10 percent bigger. As already mentioned, very distant objects have been found with z parameters up to 10 (that is, their light has been stretched about eleven times).

If we do the math and employ tangents (the opposite side divided by the adjacent side in a right triangle) we can determine the z factor from a higher-dimensional curve and compare it to what Hubble's law allows. We will see that, for all z quantities up to just over 2.0 (or three times the original), the distance to that object is farther than it is with Hubble's criteria.

THE ILLUSION OF DARK ENERGY

The same discrepancy occurs with accurately measured objects, the 1a supernovas; they are also more distant than would be expected through Hubble's law. Because of that, a new, unknown essence—dark energy—has been interposed to allow for this increase.

However, since the supernova measurements are *correct*, and those assumed from Hubble's law are simply inaccurate, there is no need for dark energy; there exists instead an actual, objective distance, and it correlates quite well with the increase seen when using tangents.

This is easy to show using our one- and two-dimensional example. However, although we use a simple demonstration of an unreal lower-dimensional world, the same finding holds for any sphere in any dimension, as all spheres, as previously described, are essentially *circles* (merely multiplied by solids of two lesser dimensions). The math can be found in the appendix for those so inclined (see Appendix B), but the results are presented below.

If we list the z values against distances assumed by Hubble's law (*Hubble/expansion*), and then against a higher-dimensional curve (*4D curve/tangent*), we find the following:

z	Hubble/expansion distance (10⁹ Lys)	4D curve/tangent distance (10⁹ Lys)	Dark energy? distance (10⁹ Lys)
0.1	1.5	4.6	+ 3.1
0.2	2.8	6.1	+ 3.3
0.4	4.6	8.0	+ 3.4
0.5	5.5	8.5	+ 3.0
0.8	7.7	9.7	+ 2.0
1.2	9.5	10.7	+ 1.2
2.0	11.5	11.6	+ 0.1
2.4	12.3	11.9	- 0.4

In all cases (up to a z parameter of about 2.0), the distances calculated by the higher-dimensional curve are greater than those assumed by Hubble's law. Hence, things with increasing red-shifts are truly farther away than Hubble's law would allow. The universe has really been larger (for the last eleven or so billion years) than previously thought; and while this discrepancy has been considered proof of another, unknown form of energy, there is in actuality no such thing. Dark energy is a "chimera" (a mystical beast found in fantasy alone); the greater distance is the *correct* distance, clearly seen when the universe is understood as a 3-sphere.

I will represent this idea as a graph to further simplify it. The distance shortfall, at each z factor, is what cosmologists have termed "dark energy." It is simply a miscalculation using an incorrect assumption (Hubble's law).

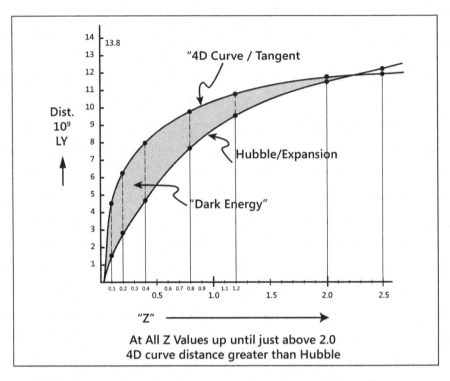

Dist. 10^9 LY

"4D Curve / Tangent

Hubble/Expansion

"Dark Energy"

"Z"

At All Z Values up until just above 2.0
4D curve distance greater than Hubble

This illustration demonstrates that the universe drawn as a 3-sphere can explain the findings of the 1a supernovas and does not require recourse to an entirely new and exotic form of energy.

ESSENTIALLY IMPOSSIBLE INFLATION

Once we adopt the 3-sphere concept, we can also do away with inflation, an entirely *ad hoc* addition without any genuine basis. Paul Steinhardt, a well-respected physicist and an early proponent of inflationary theory, now considers it almost infinitely impossible to have occurred. In an April 2011 *Scientific American* article, he notes the probability of inflation having actually happened at 1 to 10^{googol}. A googol is 10^{100}; therefore, 10^{googol} is so large a number that it would be impossible to write; hence the odds are astronomically against inflation.

His thinking is that inflation would have had to negate the second law of thermodynamics, of ever-increasing entropy, or disorder, that all things left to themselves must undergo. Furthermore, if inflation *did* occur, according to currently accepted quantum theory it would never cease, thereby allowing for an infinite number of universes and making ours nothing special. Thus, the initial reason for using it—to explain our current state—would be valueless; it would not differentiate our realm from a never-ending multitude of fellow worlds.

TRUE CAUSE OF CMB

Finally, the *raison d'être* of the Big Bang theory is CMB (cosmic microwave background); it is the cornerstone of the Big Bang's acceptance and the basis for the demise of its opponent, the Steady State theory. If there was no Big Bang, then there were no primeval infrared photons either—the supposed precursors of today's CMB. Furthermore, since we are also doing away with inflation, there must be another explanation for CMB's homogeneity. Why *is* CMB all-encompassing and the same throughout? In fact, why does it exist at all?

I propose instead that CMB is the last possible edge of our 3-sphere universe, just one Planck length less than 90^0 (after which all things disappear), stretched by the enormous z factor at such a huge tangent. Thus the final segment of the universe "visible" to us is about 10^{-35} meters (approximately one Planck length) multiplied by 10^{32}, the magnitude of the z parameter at that curvature (for a further discussion, see Appendix C). Therefore:

$$10^{-35} \, m \times 10^{32} = 10^{-3} \, m, \text{ or microwaves;}$$

and all that those microwaves, bathing our universe with back-ground radiation, entail is that edge. Since the edge or least possible size is a constant, the microwave radiation is the same throughout.

BACKGROUND IRREGULARITIES

Finally, the slight anisotropies (the irregularities of one part in 10,000) found in this background, and assumed to be secondary to vacuum fluctuations magnified by inflation, are instead the multitude of large galaxies and their clusters present throughout the universe. All parts of the universe are essentially the same: It is homogeneous. The curve just less than 90^0 away from us is similar to what is nearby. The same clustering occurs throughout. Sizes would vary, more or less, within a range of $1/10,000^{th}$ that of the entirety; thus the irregularities are a function of what exists every-where and every-when. (For more on this, see Appendix D.)

Hence, the Big Bang theory's crucial error lies in its most basic assumption. This makes it extremely difficult to root out. It is false because of the misguided use of the Doppler effect in explaining the redshift changes of distant galaxies, a very natural mistake; nevertheless, uncovering it leads ultimately to the dissolution of what has become a hodge-podge of a theory.

In the next several chapters we will endeavor to explain how the universe, as the surface of a higher-dimensional entity, can be envisaged.

Chapter 4

UNIVERSALITY OF FORCE

Centripetal attraction and granularity

EXISTENCE AND CENTRIPETAL FORCE

HOPEFULLY, THUS FAR, WE HAVE BEEN ABLE to show that the world is a surface phenomenon—a great 3-sphere on an unknowable fourth-dimensional abyss. But if this is true, how do the countless objects that constitute it make some sense to us, its inhabitants?

We must first agree on one basic assumption—*we must allow for existence*. The universe exists; this is the most fundamental of concepts, for if one does not consider existence to be real, nothing further can, or need, be considered. And if things exist, they must have shapes of some kind. And if that is true, then the entities must contain contours or edges—surfaces that enclose them.

The simplest shape is circular (it takes the least energy to maintain), and we have shown thus far that all spheres consist of circles and lesser-dimensional solids, or balls. The force exerted by such an enclosure, or sphere, is centripetal—the force toward the center. Mathematically the formula is:

$$F = mv^2/r,$$
m is the mass of the surface,
v is its speed of rotation,
r is the radius of the sphere;

and it follows that, if mass and velocity remain constant, the smaller the radius, the greater the strength or force of adherence to that surface.

EQUALITY OF FORCE

There are believed to be four fundamental forces making up our world—*gravity, the weak interaction, electromagnetism,* and *the strong nuclear force.* Gravity is by far the weakest of these. If gravity is arbitrarily considered equal to 1, then the weak interaction is about 10^{32} times as great (100 million, trillion, trillion times), electromagnetism 10^{39} times as strong (10 million times greater than the weak force), and the strong nuclear force 10^{41} times as powerful (100 times more potent than electromagnetism).

I wish to show that all of these disparate forces are really measuring the same thing—they are each different aspects of a unitary centripetal force molding all things in our world. Now, of course, gravity is felt in the *macro*-world; it holds the visible cosmos together. It grips us to the Earth; it keeps our planet revolving about the Sun; it maintains the entire solar system's orbit through the Milky Way. The other forces are much stronger, but they interact with entities in the *micro*-world.

Therefore, although gravity is much weaker, it is maintaining a much larger structure. A proton is about 10^{-15} meters; the universe, at 13.8 billion light-years, is approximately 10^{26} meters. The difference is 10^{41} orders of magnitude, equivalent to the disparity in the forces involved. Likewise, an electron is more or less 10^{-13} meters (its Compton radius or smallest possible proportion) or $1/10^{39th}$ times the size of the universe; but its internal strength—its electromagnetic force—is 10^{39} times that of the world's. Thus, strength of maintenance appears to be equivalent in these entities—it is inversely proportional to their size.

The rationale for this appears to be that, since the universe is a great 3-sphere, with protons and electrons as its surface nodes, (each presenting as a complete and intact orb), then each would be entirely (that is, spherically) encasing a similar, "infinitesimal" fourth-dimensional black hole. Each, as a result, would muster the same total counterforce (as all black holes would be essentially the same—higher-dimensional points in space, less than the Planck limit). Thus, the smaller the overall area (of that surface enclosure), the greater is the counter-pressure per unit area. Hence, surface containment, or tension must increase as the scale decreases.

BALLOON

Perhaps a simpler way to understand this is by analogy to a spherical balloon. When fully expanded, the balloon becomes a 2-sphere encasing a 3-ball of compressed air. That air is equivalent to our energy—it is the same as what exists hidden within our real, 3-sphere universe: the unknowable essence of the fourth-dimensional abyss. This compressed energy exerts an equal pressure on the entire 2-sphere, the entire spherical surface of our balloon.

A minute portion of this surface, therefore (let us say the size of a proton, or 10^{-15} meters) has to contain, or hold back, the same total energy as does a somewhat larger part (for example, an area the size of an electron, or 10^{-13} meters), for if either were to break, the entire ensemble could collapse and the energy within dissipate with a great hissing sound. So the smaller the area, the greater the effort exerted per unit size, as all aspects must maintain the integrity of the entire product.

Using the formula for centripetal force we can show that:

—

$$Force = energy/radius;$$

and since the energy we are contemplating is a constant (exerting equal pressure throughout), the force of enclosure (the surface tension) increases as the size diminishes, or, one can say, it is proportional to 1/radius. (For further discussion of this idea, see Appendix E).

We are, then, attempting to explain that, in the real world, the force that maintains a minute part (proton or electron) *or* the immense whole (universe) is linked to the total energy of the entirety divided by that object's size. Thus it is similar to our balloon with either its full surface, or any minute segment of it, holding back all of the air; if there is any puncture, the whole structure implodes. Therefore, since all energy (the essence of the higher dimension) needs containment, every particle withstands the same total pressure; the fundamental forces become one—there is a *universality of force.*

WEAK FORCE AND NEUTRINOS

If we now return to our original discussion, the last component, the weak interaction—the force that is involved with neutrons as they morph into protons and, therefore, intertwined with neutrinos (objects of minimal mass "shed" during this process)—is 1 billion times less than that of the strong; hence, the objects so defined should be 1 billion times as large, or 10^{-6} meters. Thus, if things exist (and they do) and have shapes and contours (as all things must), they have to be enclosed as spheres (possessing the least energy required), and the force of that inclusion manifests as centripetal. It is a force or surface tension that varies with scale; it decreases as size increases.

ZENO'S PARADOX AND GRANULARITY

Finally, there is understood to be a least-possible size to our universe—the Planck length (1.6×10^{-35} meters). Established at the turn of the last century by Max Planck—a founding figure in quantum theory—this size is based on fundamental concepts accepted by most, if not all, scientists. But long before Planck's seminal work, the philosopher Zeno, in ancient Greece, had already set the foundation for the concept of granularity. *Zeno's paradox*, best exemplified by the tale of a race between the swift warrior Achilles and a plodding tortoise, proves that there must be a minimal size to objects in our world.

In that parable, Zeno tries to prove that, although Achilles is much faster than the lumbering tortoise, if he were to start from behind he could never really catch up and pass. Every time Achilles is about to reach the tortoise, it moves yet a smaller distance, always keeping just ahead of the approaching Achilles. No matter how rapidly Achilles moves, the tortoise can take an extra, albeit ever tinier, step. Since this continues *ad infinitum* to ever-smaller portions, Achilles can never actually catch and pass the much slower animal.

Now, Zeno's paradox is obviously wrong. Faster objects have no difficulty catching and passing slower ones. Thus, although logically correct, it falters in the real world. The reason why is the same as what Planck showed some 2000 years later: There is a smallest scale to our world. The tortoise can only take steps down to that size; if it were to try to divide that space yet again, it could not; the space would no longer exist. Thus Zeno's paradox proves a granularity, or minimal extent to objects. If the same force that contains the universe, and the neutrino, electron, and proton, also maintains objects at Planck length, then it should be a centripetal pull somewhat more than 10^{60} times as great as gravity's, or around 10^{20} times that of the strong force.

Therefore, we find that the energy to enclose a structure of Planck proportions is about 10^{20} times as great as that engulfing a proton; or, one could say that the force of containment—the centripetal force—is that many times stronger. Thus, if things exist and are encased by spherical surfaces, the force at their boundaries is directly related to the inverse of their size: the smaller, the stronger.

What, then, are we attempting to portray? The universe, we feel, can be understood as a giant 3-sphere with its many (10^{80}) nodes each consisting of a proton, electron, and neutrino—all potential hydrogen atoms. These in turn rest on a three-dimensional backdrop of granular space composed of Planck volumes (about 10^{180} in total). All components enclose the unfathomable and energetic fourth-dimensional abyss, and the strength of that surface encasement increases as the magnitude of containment decreases.

Chapter 5

WITHIN AND BEYOND

*The higher dimension as both within
and without all that exists*

CENTER AND EDGE

E HAVE CONCLUDED THUS FAR that our universe is a 3-sphere of immense magnitude, held together by the force of gravity; and that the centripetal basis of this force is found in all quanta and is related inversely to their size. Let us, therefore, bring up some other interesting and unusual concepts.

The edge of the universe appears to be equally distant from all points. This is an offshoot of the Copernican or cosmological principle that the universe is the same in all directions, from any point of view or any position in space. Thus, there can be no absolute center, merely an edge which we visualize as a beginning in time (if we accept the Big Bang scenario) occurring some 13.8 billion years ago. From any point to this edge of time, the world's radius measures about 10^{26} meters (13.8 billion light-years).

PLANAR REALM

But what does a structure actually look like if every point is at a center equidistant to an edge? Theoretically, it should resemble the surface of a globe, a 2-sphere, a surface with an apparent edge, from any point, in any direction, at 90^0. Beyond that, there would be nothing, as no further electromagnetic wave or information could be obtained; all would be infinitely far away.

The easiest way to picture this *should* then be by using a simplified two-dimensional model. Thus, instead of describing the universe as it actually appears, let us assume that we are *two-dimensional* inhabitants of a seemingly flat world that, in reality, is encircling a three-dimensional orb (a 2-sphere surface of a 3-ball). Since we would think that our world is on an endless flat plane (extending infinitely north–south and east–west) though, in actuality, it rests on a curved surface, what would we see? If we drew this in a simplified manner we could get the following:

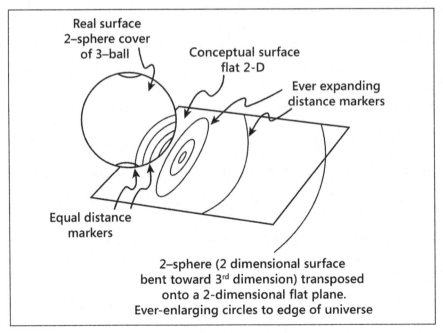

Real surface
2–sphere cover
of 3–ball

Conceptual surface
flat 2-D

Ever expanding
distance markers

Equal distance
markers

2–sphere (2 dimensional surface
bent toward 3rd dimension) transposed
onto a 2-dimensional flat plane.
Ever-enlarging circles to edge of universe

The picture is of ever-enlarging circles out to infinity. We are at the point on the sphere that attaches to the supposed flat surface. As we look back onto it, we visualize a flattened plane with concentric, ever-enlarging circles from the very small (which is us) to the extraordinarily large (or the edge of the world). We are really seeing equidistant circular markers that appear farther and farther apart due to the distortion caused by our viewing on a curve. As we further examine the space within our own minuscule spot, we get ever-tighter circles, closer and closer together, finally reaching a point of smallest possible size (Planck scale). We could not go smaller, for if we did we would no longer be in our world; we would, instead, be entering the higher, or third dimension.

PROTON OBSERVER

Therefore, let's assume that the smallest perceptible or tangible object in our world is us, the viewer; however, the tiniest possible size is really many times smaller—the Planck length. If we change our story somewhat and make the viewer a proton, we have what represents the smallest object of permanence in our universe, but still many times greater than true granularity. From the proton's perspective, then, the world goes outward in ever-greater concentric circles to its very edge. Surrounding our proton is the next tangible—but not nearly as massive or dense—permanent structure, the electron. Surrounding that, in the great distance, is the ephemeral, barely discernible neutrino; after that, way beyond all else, lies the very edge of the universe itself.

ONE DIMENSION MORE, ONE DIMENSION LESS

If we change this analogy a bit, we as real, three-dimensional beings can easily see that the entire two-dimensional universe is suspended in what, to its inhabitants, is unfathomable: third-dimensional space. The flat world of two dimensions is surrounded by an infinitely larger third dimension. In reality, a universe of any number of dimensions, in order to truly exist, must be placed within a higher-dimensional venue. A one-dimensional world, a straight line, needs to be drawn on a two-dimensional surface. A two-dimensional plane, in turn, has to lie within a third-dimensional enclosure. Our three-dimensional world, therefore, would require a fourth-dimensional milieu in order to actually exist.

Furthermore, an inhabitant in any universe, of any-dimensional size, really interacts with objects one dimension less than what that world contains. Thus, although the world is situated in a higher expanse, that being would interact with objects from a lower realm. So a one-dimensional denizen, a segment of a line, only "sees" points. But by interacting or exerting force against these points, it can establish the true extent or heft of other fragments. Likewise, a two-dimensional individual only visualizes lines; however, by pushing against them or observing them from differing aspects, that person can establish that these lines may enclose other material or mass.

HOLOGRAMS AND THREE-DIMENSIONALITY

In our three-dimensional world, we "see" only surfaces. However, we can interpret them as the covering to more massive internal structures by moving about them and touching or pushing. This conclusion, then—that we only interact or gain information from surfaces—forms the basis for the *holographic universe*, a belief that, in our world, all information is obtained from two-dimensional surfaces. Therefore, although our real world appears three-dimensional, it exists in fourth-dimensional space but information is assessed two-dimensionally. In a similar vein, our two-dimensional inhabitant's universe is enveloped within three dimensions, but its individuals only interact with one-dimensional lines.

In reality, there are only three dimensions; one- and two-dimensional worlds are mere mathematical constructs. All things in our universe must employ the same three directions: width, length, and height. Even a simple, straight line, once drawn, must contain all three. Similarly, there can be no truly flat surfaces that fully lack height. So, the examples of lower dimensions are strictly imaginary; but they can be quite useful in helping us to visualize the higher-dimensional realm.

WITHIN AND WITHOUT

If we now resume our visit to our make-believe, lower-dimensional world, although our being from two dimensions, our flat proton observer, is immersed in three-dimensional space, it is unaware of this fact; its world does not allow for that third direction, height. Thus, to that proton, there is no outside (no up or down); there exist only smaller and smaller spaces until the Planck scale, and nothing beyond. To that proton, the next dimension can only be conceived of as within itself, as an ongoing or infinitely ever-smaller division of "space" continuing beyond what is possible in its planar world.

So if we return to our "real" three-dimensional world, the actual proton observer (the smallest material object) considers that a higher or fourth dimension could only exist within the three it understands. We, however, assuming the "vaunted" perspective of a higher-dimensional "being," know that the fourth dimension, although considered to be within, is in fact beyond all that exists. To the proton, "within" means farther inside itself, within the nucleus of a hydrogen atom. Thus, the proton observer may understand that a higher dimension exists but can only visualize it as situated inside what is already present. Since the proton is the center of a hydrogen atom, the other dimension is ever-farther within that center—infinitely divisible.

—

NUCLEUS, BOTH CENTER AND EDGE

Hydrogen, the simplest element, with a single proton, makes up about 75 percent of the mass of the world; helium, with a nucleus of two protons and two neutrons, constitutes most of the other 25 percent. Yet the remaining elements, with multiple amalgams of protons and neutrons, combine to account for much of what we consider to be existence. So, within nuclei, there are two basic entities, protons and neutrons, but, at the same time, although every nucleus is a center of the universe, each also becomes an edge to our world.

What, then, are we trying to describe when we say that there is no absolute center to our world, only edges? We are picturing a spherical object—a 3-sphere—whose edge in *time* is always 13.8 billion light-years away (the beginning of time to a Big Bang advocate), yet whose edge in *space* is to be found within the very centers of all atoms. Thus, every spot, every nuclear core, is at the center of the world yet at the same time its edge.

Chapter 6

THE 3-SPHERE

*Attempting to visualize
what is beyond our grasp*

FIRST ATTEMPT AT A 3-SPHERE

I F WE ARE IN AGREEMENT that our world is a 3-sphere covering some unfathomable 4-ball or fourth-dimensional object, what does the actual 3-sphere resemble? It is really impossible to draw a higher-dimensional object, since the space it occupies does not exist in our world? If we use lower-dimensional examples, we can, at least, begin to imagine what such an object would look like if it were only three-dimensional.

Again, as shown in the previous chapter, if our world were only two-dimensional—if it were, that is to say, a 2-sphere covering a three-dimensional orb—and if its inhabitants did not know of this higher-dimensional curve, then they would see things as circles within ever-enlarging circles stretching to infinity:

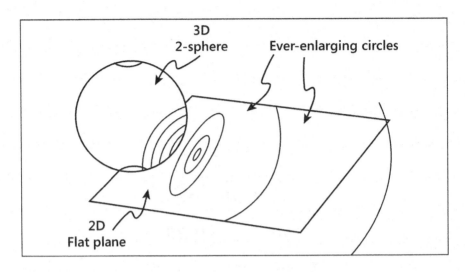

If we now transpose this to our three-dimensional world by adding height, what a two-dimensional inhabitant considers evermore distant concentric circles becomes, to us, ever-increasing concentric spheres:

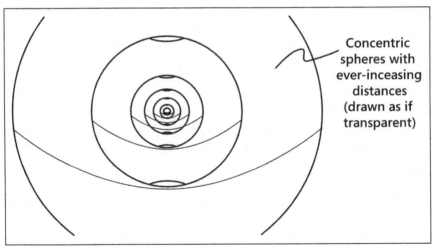

But a two-dimensional inhabitant, if knowledgeable, could construct the *semblance* of a three-dimensional object (a 2-sphere) by taking each circle and placing it next to its larger neighbor. A drawing in two dimensions could be the following:

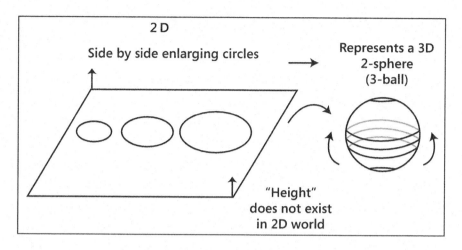

However, since we know that the two-dimensional inhabitant cannot *conceive* of height, that individual could only draw it within his or her own flat world. In a similar manner, if *we* see ever-enlarging spheres, we could place them side-by-side to begin to establish a true 3-sphere, even though our construct cannot actually describe the higher dimension, since that space does not appear to exist in our world. If we were to draw it we would get the following:

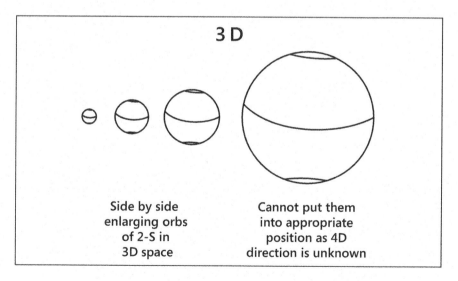

FURTHER ATTEMPT AT A 3-SPHERE

Now, in a two-dimensional world, there would not be an absolute, single "central" proton but a huge multitude of them dispersed throughout a great, flat planar surface, each of which would in actuality be a depiction of the 2-sphere surface at that point in space. So all would look the same, only spread throughout that world. Also, as previously noted, according to the Copernican principle, all would be at the center of that universe, equidistant to its edge. Since, moreover, they would all be part of the same universe, the largest possible circle, the universe itself, would be centrally positioned, and the protons would be disjointed and randomly placed "nodes." Let me picture it (for the sake of drawing, we are omitting all spherical matter between the neutrino—10^{-6} meters—and the universe—10^{26} meters—obviously quite a lot).

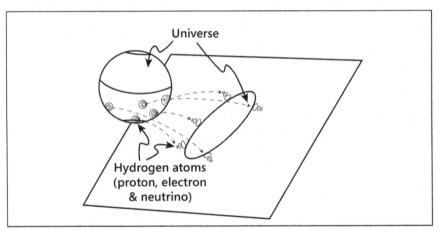

A 3-sphere, the cover in three dimensions to a higher-dimensional object, can similarly be depicted. If we wish to redraw this same concept, we can use one "great" 2-sphere as a center with nodes in all directions ending as protons. These, then, become the surface of that 3-sphere and are found randomly all over three-dimensional space, as illustrated on the next page.

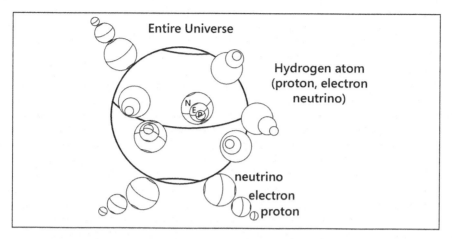

Entire Universe

Hydrogen atom
(proton, electron
neutrino)

neutrino
electron
proton

FINAL ATTEMPT

What we have shown is a central, largest possible 2-sphere
(the "infinitely" large 2-sphere universe) with nodes in all directions ending in tangible protons, each as part of a surface. Since
the 3-sphere, our world, presents to us as a volume, to find its full
extent, its radius would be cubed ($2\pi^2r^3$); however, its own surface,
similar to that of any volume, would be one dimension less, or
squared ($4\pi r^2$). Therefore, if the universe is about 10^{26} meters,
but a proton is around 10^{-15} meters, the difference is 10^{41} times,
and the surface therefore contains, more or less, 10^{82} protons.
(For rounding-down purposes, we will use a proton as $1/10^{40th}$ the
size, not $1/10^{41st}$ that of the universe. Hence the surface, the two-
dimensional skin of our three-dimensional 3-sphere, should contain somewhere nearer to 10^{80} protons.)

Now, our world is believed to contain approximately 10^{80} protons; consequently, we are showing it to be a 3-sphere with nodes
(hydrogen atoms), making up its surface, distributed throughout
three-dimensional space. If we now wish to show a simpler concept, we can take a great portion of our world (a giant cube of
space with its many protons) and reshape it by bending and placing it onto a sphere:

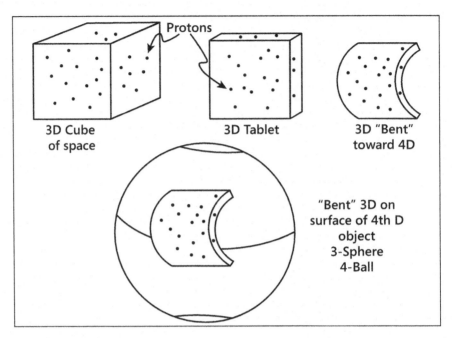

The result is a 3-sphere with its proton surfaces. We can re-draw it as follows:

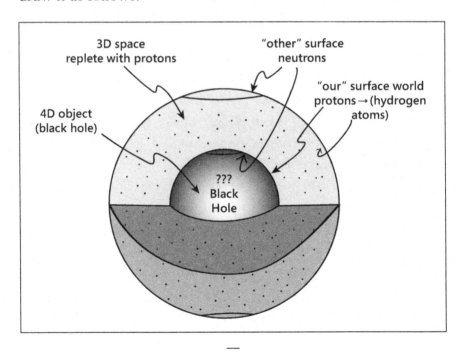

"OUR" SIDE AND THE "OTHER" SIDE

We now have a 3-sphere or cover to a fourth-dimensional entity, drawn in three-dimensional space, with its surface, "our" world, displayed as hydrogen atoms (protons covered by electrons and distant neutrinos). However, as the 3-sphere is *itself* a surface (albeit in fourth-dimensional space), it must also have two sides. So "our" side, the side we acknowledge as within our world, would contain protons as its edge; and the "other" side, the side facing the fourth-dimensional abyss, would contain neutrons, the hydrogen atoms' doppelgängers. In the next chapter, therefore, let us attempt to understand what—in the world, or outside it—is a *neutron?*

Chapter 7

NEUTRONS—THE "OTHER" SIDE

*The two sides to the surface
or 3-sphere we call our world*

4D MIRROR IMAGE

F PROTONS, WITH THEIR ASSOCIATED electron and neutrino covers, constitute the final nodes of a giant 3-sphere (the many hydrogen atoms of our world), what do the neutrons (the other nuclei inhabitants) signify? We will, hopefully, show that the surface to our world really has two sides (albeit only visualizable in higher, fourth-dimensional space); therefore, the neutron represents a fourth-dimensional "mirror image" of the hydrogen atom. It is the 3-sphere node on the "other" side of that surface.

We have, I hope, shown that each nucleus is the equivalent to a minute piece of the surface of the universe; hence, all that is contained within a nucleus becomes the world's surface material. In hydrogen there is simply a proton; however, in all other elements the proton coexists with its fellow neutrons. Therefore, just as each proton is at the center and, simultaneously, the edge of all that exists, the neutron is the same, only inversely.

BETA RADIATION

The neutron becomes a hydrogen atom, but to us it is inside-out. Its own center, the equivalent of a proton, becomes its exterior; its electron cloud is found *within*, and deeper yet, its neutrino. Hence, in *beta radiation* when, for whatever reason, a neutron crosses over from the "other" to "our" side, it becomes a proton, an electron, and a distant covering sheath or neutrino (actually, an antineutrino). Similarly, when a proton crosses over to the "other" side, it takes its electron (actually, an antielectron, or positron) and neutrino and, when fully stabilized, becomes a neutron. So the act of crossing over appears to be a movement that inverts all parts and places the internal ones externally, and vice-versa.

We can easily demonstrate that, from a fourth-dimensional perspective, the inside and outside of a three-dimensional body are equally visible. To do so, let us use a two-dimensional example. In such a world, an embedded observer (an eye) can only see the outside of another two-dimensional entity. It must pierce that surface to view within:

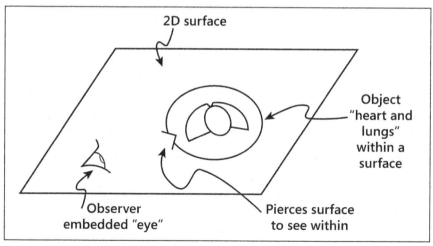

The observer merely sees the surface (or "skin"); the entity's internal aspects (or "heart and lungs") can only be found by opening it up. However, to a three-dimensional person, both the inside and outside are visible at one time:

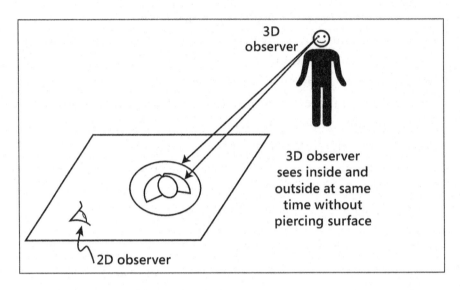

CROSSING OVER

Thus, to a higher-dimensional being, our world would be transparent. However, the crossing over of a neutron to a proton, or proton to a neutron, is more than just viewing all aspects at the same time. It is, in actuality, a crossing over to "our" side from the "other" side, or vice-versa. All the surfaces are reversed: Whatever was in goes out; whatever was out comes in. And, importantly, the analogy of seeing three-dimensional objects as transparent does not adequately explain this process. The object really turns itself inside out when it crosses over and is visualized from a higher perspective. What is really occurring is that the two sides of the surface that separates three from four dimensions are crossed. A depiction of how this process could occur appears on the next page.

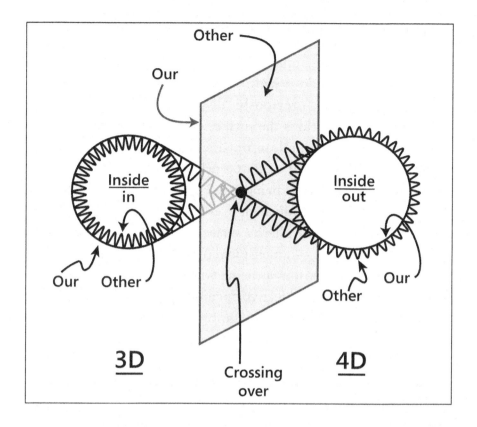

The sphere (proton) with "our" surface as part of three-dimensional space, and the "other" surface as virtual only, crosses over by going into the higher dimension—it inverts; thus, the virtual side now becomes the outside, and the real side becomes the inside. Therefore, the act of crossing over changes real into virtual, and vice-versa. The virtual entity, the virtual hydrogen atom or neutron, has its inside on the outside, its outside on the inside. It has completely reversed. Although it is the same size as a proton (the smallest object perceptible in our world), it actually is a reversed proton (or antiproton) containing a reversed electron (or antielectron) and a reversed neutrino (or antineutrino) tucked within; the bigger a particle was in our world, the deeper and smaller it becomes in the virtual realm.

COMPLETE REVERSAL: POSITION AND SIZE

If we once again employ a lower-dimensional example, perhaps we can show *why* a neutron would be the inverse of a hydrogen atom. As previously discussed, our make-believe two-dimensional world is the surface of a 2-sphere. It has an inside (the real aspect) and an outside (the imaginary one). Although, to an inhabitant of *that* world, only what is real exists, we as higher-dimensional beings are cognizant of *both*. Thus we can see that there are equivalent ever-enlarging concentric circles on both sides of the 2-sphere's surface—one side real, the other imaginary. Now, a two-dimensional person can only visualize the higher, or third, dimension as being within the minute confines of a nucleus, the smallest objective opening possible. So the entirety of the imaginary hydrogen atom (antiproton, antielectron and antineutrino) is viewable only, since it represents this "other" realm, through a vista the size of a proton.

If we now reenter our real, three-dimensional world we will see the same changes occurring, only with the addition of a third direction, height. We can show that the larger an object is, the farther away it would appear when reversed; therefore, the diminutive, compact, antiproton core becomes the exterior, the distant, ephemeral antineutrino its miniscule center, and the middling antielectron remains somewhere between the two (see next page).

All elements, with the exception of hydrogen, consist of both protons and neutrons. These particles—or nucleons—represent "our" and the "other" side, the surfaces to the many (10^{80}) nodes of our giant 3-sphere universe. However, since the proton exists in the real world, it is surrounded by an electron and a neutrino; the neutron, on the other hand, is found in the imaginary realm. It is, in a sense, the mirror of the real; it is the inverse, observable only through the distorting lens of a nucleus.

—

50

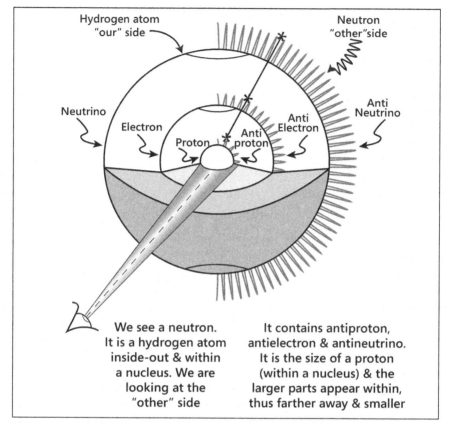

Hydrogen atom "our" side

Neutron "other" side

Neutrino

Anti Neutrino

Electron

Anti Electron

Proton

Anti proton

We see a neutron. It is a hydrogen atom inside-out & within a nucleus. We are looking at the "other" side

It contains antiproton, antielectron & antineutrino. It is the size of a proton (within a nucleus) & the larger parts appear within, thus farther away & smaller

TWO-SIDED COIN

One final example may, perhaps, be helpful in clearing this up. If we were to return to the two-dimensional world, we could use a fictitious coin with heads on the real side and tails on the imaginary side. Inhabitants of that world would only recognize heads; tails would not exist. However, to a three-dimensional person, to us, this extremely thin coin (one Planck length in thickness) has *two equally distinguishable surfaces*. We need to look "under" the two-dimensional plane to find the other aspect, but both exist. In a similar manner, our three-dimensional object—a proton—has two segments; a real and an imaginary one. We recognize only what is real; but a fourth-dimensional observer, by looking "under" (an almost incomprehensible direction in our world, best understood as "within") can acknowledge both the real and the imaginary parts.

DOPPELGÄNGER

It is not merely that our *macro*-world would be transparent to a higher-dimensional being; objects in our *micro*-realm would *also* present to that individual with their doppelgängers—the same, only inside out. This is what a neutron really signifies. The neutron is the inverse of the complete hydrogen atom; it is the same, only on the "other" side; it is inside out. It represents an antiproton, surrounding an antielectron, encasing an antineutrino. Its total size is that of a nucleon, or antiproton, and the other (anti-) aspects are found within, in an ever-smaller perspective. The neutron is our world turned on itself and placed inside a nucleus.

This explains why, when it emerges to our side—either exiting the nucleus, or staying put but changing from a neutron to a proton—we find its innards spewing outward. It is also why the neutron is *neutral*. It has both an antiproton and an antielectron, and they counter one another. Finally, this is the basis of all antiparticles. We will next show why they *are* so rare, and why they occur only in violent collisions.

Chapter 8

ANTIMATTER

The same, only inside out

DEPICTING A NEUTRON

I HAVE BEEN TRYING TO SHOW why the neutron is on the "other" side of the surface that encases our universe. Since we only can visualize this other surface as an ever-decreasing center, neutrons are found within the smallest stable object, the nucleon. They are the size of protons yet contain within themselves a diminutive electron and a much smaller neutrino. Finally, since they are on the other side, they would be the equivalent of similar virtual entities, or antiparticles. So if we draw a hydrogen atom, the neutron would look as follows:

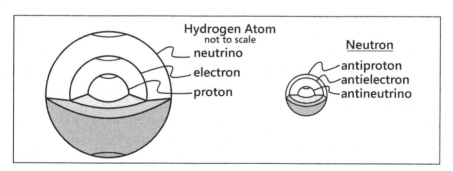

ANTIPARTICLES

Antiparticles are exactly the same as particles, except reversed. Their charges are opposite, but in other respects, size and mass, they are equal to particles. They are really the "other" side to particles, the virtual mirror images, not seen unless, due to a collision of some kind (cosmic ray, particle accelerator), an entity has been turned inside out (twisted or pushed onto that side, allowing its antiparticle to enter our world). Therefore, they are rarely found, since most particles are stable and present their normal facades in our three-dimensional world.

When a particle and its antiparticle collide, they both lose their coherence; they no longer exist as a surface to our universe. The entity (particle–antiparticle pair) becomes a *virtual object*, lost in the vastness of space, a volume 10^{41} (100 thousand, trillion, trillion, trillion) times as great as the surface (for a further discussion of this idea, refer to Appendix F). That surface is then reestablished, and the energy of its closure is the centripetal force of gravity: $F = mv^2/r$. Since we are discussing the spin of the entire universe through time, the velocity (v) is that of light (c). So it can be written as: $F = mc^2/r$, which, if cross multiplied, becomes: $rF = mc^2$. Since mc^2 is the energy of a particle, the force of closure is equivalent to the total energy in the pair of particles, and they disappear as mass, becoming, instead, what appear to be high-frequency energetic photons.

SPIN

Since the spin of the entire universe presents as a physical movement within the fourth dimension or time, so too would be the motion of its many parts. Another perspective on particles and their antiparticles envisages particles as *continuously spinning entities*—going into and out of our understandable world:

—

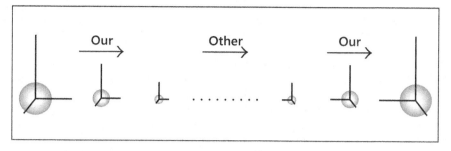

The particles "spin" by constantly leaving and reentering the real world; but, as already drawn, there is a crossing-over when this occurs:

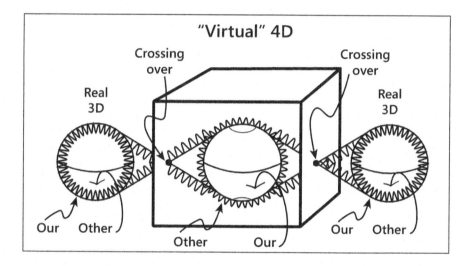

Hence, the *quantum* world, the world of the very small, is the surface of *our* world. Spin is the motion of that surface as it maintains centripetal force; therefore, all particles, since they are segments of that surface, must spin. It is an imaginary motion in the sense that it brings a particle into the fourth-dimension, the "other" side of the universe. However, since all particles partake of this motion, all particles constantly enter and exit this virtual realm.

Chapter 9

THE STANDARD MODEL

Its basic problem

CURRENT CONCEPTS

I HAVE BEEN ATTEMPTING TO SHOW, thus far, that the universe is a great 3-sphere, and that the Big Bang never happened. But if there was no initiating event, then much of today's thinking concerning particle formation—the *Standard Model*—also needs to be recast.

This model bases the formation and evolution of particles on an initial "cataclysm" leading, subsequently, to expansion and cooling. As things became ever-less energetic, different entities condensed out. Initially, at extremely high temperatures, particle–antiparticle collisions caused annihilations, leading to high-frequency photons, which, in turn, readily reverted to their original entities. These processes got continuously less frequent as the overall temperature fell, leaving us, finally, with today's particles in a cold, slowly expanding world.

Although this scenario may appear plausible, since there *was* no primary event, the concept of expansion and cooling underlying it cannot be correct.

I am proposing that the stable, final particles—the protons with their accompanying electrons and neutrinos—are actually the "nodes" of a giant 3-sphere, our universe. The protons are its actual surface; hence, there should be 10^{80} in total. Wherever there is a proton, there is a larger but much less dense electron cloud surrounding and neutralizing it. The neutrino is merely an evanescent, distant veil to this union. The neutron, however, though it presents as the entire combination, is on the "other" side, within the smallest of centers (or nuclei), and inverted. Thus it contains the anti-selves of these particles, inside out.

BARYOGENESIS

Although these concepts of "within" and "inverse" may seem bizarre, they are no more so than the Standard Model. In order for today's accepted beliefs to hold, we must allow for an entirely magical occurrence—*baryogenesis*—the baffling disappearance of, essentially, the world's entire antimatter. Current scientific ideas are founded on readily reproducible particle–antiparticle annihilations leaving highly energetic photons, which, in turn, easily revert back to particle–antiparticle pairs. These reactions have been produced in high-energy collisions obtainable with current accelerators.

But the Standard Model starts, as its theoretical basis, with an equal number of particles and their twin antiparticles. Why, during the universe's cooling and condensation, do only particles survive? If the Standard Model is correct, this equality should lead to a complete annihilation of both—of all matter—leaving a world solely composed of energy, a massless sea of photons.

Baryogenesis, an unknown process that saves a very small per-

—

centage of *particles* (about one per hundred million) while destroying practically all *antiparticles*, has been inexplicably raised as a solution. From out of some magician's hat a mysterious phenomenon unfolds that permits this slight discrepancy; and the minute advantage leads to our current world, consisting almost entirely of matter.

Baryogenesis, therefore, just like inflation, is a wholly contrived device. It establishes an end result, our universe, without providing a relevant cause.

LEPTOGENESIS AND NEUTRALITY

Finally, for baryogenesis to succeed as a mechanism, an equally improbable but exactly analogous process—*leptogenesis*—must *also* have occurred. Baryogenesis led to the existence of baryons—protons and neutrons. Leptogenesis does exactly the same with electrons, allowing for their salvation through the demise of their twins, antielectrons (or positrons). Neither of these processes is even vaguely understood; and since they did not occur simultaneously but in separate and distinct cooling phases, why, even if they did take place, would they have led to a precise equality in their final products? Why should there be an electron for every proton? Why should the world appear to be neutral?

Current theory, therefore, is entirely *ad hoc*. It is decreed to have occurred, because it is obviously needed, but no rational explanation for it has been proposed. If we allow, instead, that the world is really a 3-sphere replete with a great multitude of nodes (10^{80}), then each would represent a hydrogen atom. Each would present as a surface appendage, as a proton, with an electron-neutralizing cloud and a distant, barely discernible, neutrino. Finally, the neutron, the same node only on the "other" side (hence, from our perspective, inside out and confined to a nucleus), can also be accounted for.

HYDROGEN/HELIUM RATIO

The *raison d'être* for the Standard Model has always been its explanation of the hydrogen–helium ratio. This proportion, by mass, of hydrogen to helium (75–25) could not be accounted for solely by fusion in stars, given only 13.8 billion years (the time allowed since the supposed start of the world).

But I am maintaining that there *is* no understandable start to our world; its beginning is beyond comprehension. It has been moving through time, the fourth direction, perhaps forever. In this essentially unlimited time, hydrogen, in innumerable stellar furnaces, could have fused into helium, leading to the current ratio. Therefore, although the Standard Model may be able to explain this ratio through an initial cataclysmic event, the basic assumption of a Big Bang is faulty, and the conclusion is unacceptable.

Thus far, we have been establishing a basis for the fundamental, stable particles—protons, electrons, neutrinos, and neutrons; in the next part of the book, we will attempt to explain some of their inherent strangeness.

PART TWO

Quantum Queries

Chapter 10

WEIRDNESS

*Making sense of
the strange quantum world*

CONCERNS

FULLY COMPREHENDING THE QUANTUM WORLD is probably impossible when you only consider it from our three-dimensional realm. Although it can be explained in mathematical terms, a satisfying description of its actual nature has never come into focus. The title of this book gives an idea of the paradox involved. It has been taught—literally drilled into students—that if one can explain it, one does not understand it.

This book tries to give some meaning to a very strange subject. We have shown, so far, that this is best done from a higher-dimensional approach. We have established that this other realm actually exists but is unfathomable to us solely in three dimensions. The cosmic redshift is proof of a curved, higher-dimensional universe, a 3-sphere. Its visualization (in lower-dimensional terms) leads to a concept of surface nodes, or nuclei; these, in turn, explain the lack of antimatter, the equal number of protons and electrons (and neutrinos), and the real basis of neutrons.

Using this same framework, a rudimentary understanding of the quantum realm becomes possible.

There are quite a few unusual aspects of today's quantum world that call for further clarification. These include at least the following: *wave–particle duality, zitterbewegung,* the *Pauli exclusion principle, spin 1/2, entanglement* or what Einstein termed *"spooky" action at a distance,* and the *double-slit experiment.*

The first several we will attempt to explain from the basic perspective of spin in a higher dimension. Entanglement, I think, can best be understood from a somewhat different vantage point. Finally, we can hopefully shed some "light" on the always elusive double-slit experiment.

WAVE–PARTICLE DUALITY

All entities at the quantum level are at the same time wavelike and particulate. This is not conceivable in our macro-world, a place of supposedly solid objects. Things in our world present as real and tangible; they have acknowledged shapes and surfaces. They are what they are, or at least we think that is so. In the micro-world an object can be described equally as a wave *or* a particle. There is no hard distinction, merely a fuzzy sameness.

I wish to show that, although a particle (a proton, for example) is a real entity, it also has an imaginary component, since it exists in both our world and in a virtual realm (separated by a fourth-dimensional divide). Therefore, it comes and goes; it spins forever, into and out of this other setting. We have shown this behavior, or spin, in several illustrations; but it remains quite difficult to adequately explain. We will, therefore, use a lower, two-dimensional comparison to make it more accessible.

Let us pretend that, in this fictional, lower-dimensional realm,

—

the elementary particles are 2-spheres (the surfaces of what, to us, would be ordinary globes). These entities continuously spin in what would be unknowable three-dimensional space, and their reflections comprise the two-dimensional planar world. If we were to give these spinning 2-spheres both a real (bright, fully visible) and a virtual (dark, invisible) surface, they would appear, thus, to come and go. In many ways this rotation would mimic the phases of our Moon, which, due to the peculiarities of its orbit, has one side constantly facing Earth, the other forever hidden from view. Therefore, to us, the observable half slowly alters its appearance, changing, on a more-or-less monthly basis, from full, to partial, to nothing, and then back again to full. It can be represented like this:

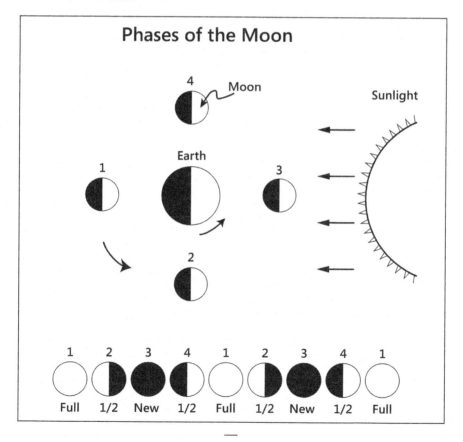

Our particles (2-spheres rotating in three-dimensional space but visualized by our two-dimensional inhabitants as part of their landscape) act similarly to our image of the Moon; they will come and go; therefore, we get the following:

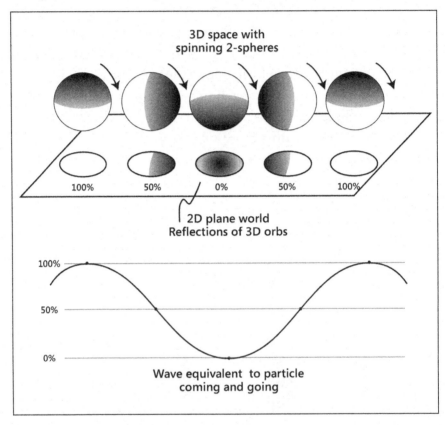

The particle envisioned as real (the reflection of the 2-sphere onto the planar surface) goes from full, to one-half, to none, and finally back, once more, to full. Since the actual particle in three-dimensional space constantly spins, the two-dimensional inhabitants see a reflection that comes and goes; to them it is a particle that resembles a wave.

We now transpose this image onto our real, three-dimensional world. Here, the particle is spinning into and out of our understandable space, changing constantly (getting smaller until it disappears, then getting larger until full-sized). Thus, all surface particles of the immense 3-sphere, our universe, continuously spin into and out of view; all present as wavelike and particulate simultaneously.

ZITTERBEWEGUNG

Another unusual aspect of quantum theory is, in German, termed *zitterbewegung*, or "jittery motion." All particles seem to undergo this very rapid movement from side to side. If we apply the same concept to it, we can show that this jitter represents the constant disappearance of the particle, on the one side, and its reappearance on the other:

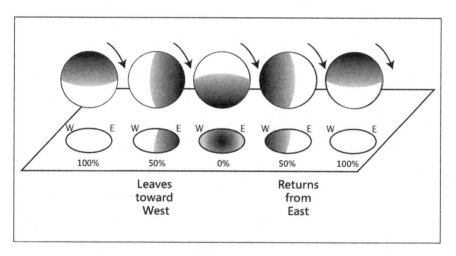

The particle, in our example, continuously departs toward the west and returns from the east; hence it is on one side then the next—it jitters back and forth. The speed of this process has to do with its spinning motion at c (the speed of light). Since the entire universe endlessly travels toward the future at this constant velocity, each particle must also similarly move.

If we take an electron whose circumference, or Compton

wavelength (its smallest size; less than that and it compresses to a black hole) is approximately 4×10^{-13} meters, and whose velocity of spin is 3×10^8 meters/second (the speed of light), it would therefore come and go approximately 7.5×10^{20} (or 750 million, trillion) times per second. The math requires dividing velocity by size. (For a fuller discussion, see Appendix G.) So the unusual findings of wave–particle duality and zitterbewegung can be understood by picturing spin into, and out of, a higher plane.

PAULI EXCLUSION PRINCIPLE

We next turn to the Pauli exclusion principle, named for Wolfgang Pauli, a brilliant twentieth-century physicist. One of his many significant accomplishments was to establish that, in order to coexist, like particles of matter—*fermions*—must have opposite spins. The mathematics is quite involved, much too difficult for this book, but the idea can be readily visualized.

Again, we return to our two-dimensional world. When two spinning orbs come into contact, they can coexist only if they are moving in the same direction:

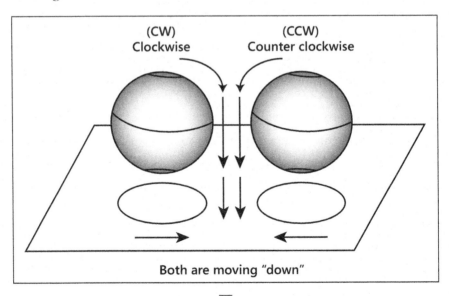

When a pair of adjacent 2-spheres spins *oppositely* (one clockwise, the other counterclockwise) they can coexist, since, upon touching, they would be moving in the same direction—"down." However, when both spin *concurrently* (either clockwise or counterclockwise) they cannot, since, upon contact, they would disrupt one another, one "up," the other "down":

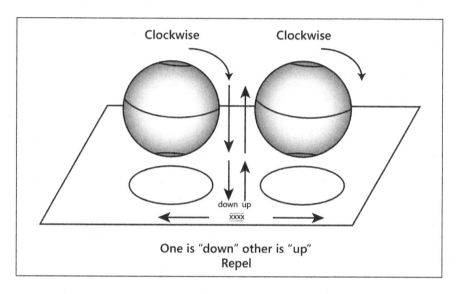

To those inhabiting the two-dimensional universe, the spin is not a real motion, it is "other-worldly"; but contact becomes impossible when one goes "up" and the other "down" (whatever those directions, in that world, actually signify). Thus, through this lower-dimensional analogy, some basic but unusual aspects of the quantum world become comprehensible.

SPIN 1/2

Consider one further unusual finding of quantum theory—spin 1/2. The universe of particles is divided into fermions (already mentioned) and bosons. Fermions, named for Enrico Fermi, an important Italian physicist, are particles of *matter*; bosons, so-called after the influential Indian scientist Satyendra N. Bose, are carriers of *force*. An example of a fermion is an electron; that of a boson is a photon.

When a fermion spins, it must make two complete turns (720^0) to get back to its original position; bosons, like all things in the real macro-world need rotate just once (360^0). This extra trip around obviously does not make sense in ordinary three-dimensional space. Why should something rotate two times to return to where it started? If we see it through a higher realm, this weird aspect can become somewhat understandable.

Again, as before, we use a lower-dimensional analogy. We will populate our make-believe two-dimensional plane with inhabitants that look just like us; however, since everything is flat, they are only aware of their front sides. Their backs are on the other aspect of this surface—similar to our prior example of a very thin coin. To these inhabitants, their backs do not exist. To complete the image, we place a special mirror in front of them. As they turn about, they fully disappear (in two-dimensional space); however, in this weird mirror (situated in three-dimensional space) we are still able to see their frontal aspects, only reversed (right would become left, and vice-versa). They have literally *traveled through themselves*; they have completely inverted. It is exactly analogous to our previous examples of neutrons and antimatter. If we in the real, three-dimensional world were to count the number of complete turns (how frequently their faces appeared), we would see two full rotations; however, to the inhabitants, there is simply one (as illustrated on the next page).

Therefore, a fermion (an electron), an object that is higher-dimensional, actually turns two times (albeit, once in the real world and once in the virtual realm) in order to make one full rotation. We will, however, show in a later chapter that a boson—a photon—because it is strictly three-dimensional, acts in a manner sim-

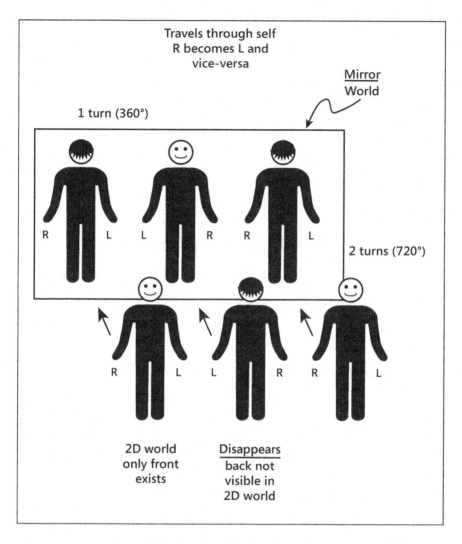

ilar to other merely "real," or macro-objects; it turns only once (it does not spin into and out of a higher plane). We will also illustrate that, because it is real, not imaginary, it does not come with an antiphoton twin, nor does it exhibit *zitterbewegung* or take part in the Pauli exclusion principle. Finally, we will demonstrate that what has always been considered wave-particle duality for our boson (photon) is an entirely different concept than for a fermion.

Chapter 11

NEWTON AND EINSTEIN

Newton's instantaneous action versus Einstein's unreachable velocity: Are they the same?

GRANULAR UNIVERSE

I HAVE BEEN DESCRIBING SOME of the truly unusual aspects of the quantum world. I have used fictional two-dimensional analogies to show how spin leads to the coming and going, or wavelike structure, of particles. However, to explain entanglement, or "spooky" action at a distance, one needs to use an entirely different concept—the idea of a granular universe.

Max Planck established, in 1900, that there was a smallest size to our world, since known, in his honor, as the *Planck scale* (length and time). However, we have shown that this concept predates, by some 2000 years, today's science; it was originally set forth by the ancient Greeks under the guise of Zeno's paradox. Zeno, and his followers, were attempting to prove that motion was an illusion. They reasoned that, before a person could travel any distance, he or she had to travel one-half of the way; yet, before getting to that halfway point, that same traveler had to move one-half of that now-shorter distance, and so on, *ad infinitum*. Thus, if the world were infinitely divisible, there would be no motion, as one would need to take an infinite number of steps to get anywhere.

Obviously, this is incorrect. The world is in a constant state of flux; all things move. Thus, Zeno's paradox, by being mistaken, demonstrates that there cannot be an infinite divisibility in our world; it shows that there must be a smallest size to things—the Planck scale.

MOVEMENT THROUGH TIME

I have been saying that the world is a great 3-sphere, moving forward through a higher dimension. We understand this motion as time. It is unidirectional, going only into the future; and it is constant. It is every Planck moment since (if one believes in the Big Bang) the beginning of our universe; or, more correctly, every moment to just less than 90^0 on a great curve—our 3-sphere.

Each Planck moment lasts 5.4×10^{-44} seconds and, since the universe has existed, or can be extrapolated back, for 13.8 billion years (or 4.35×10^{17} seconds), there would be about 8×10^{60} such instances (4.35×10^{17} s / 5.4×10^{-44} s). Thus, the universe moves continuously, but in a stop-and-go fashion, always forward in time—Planck moment to Planck moment. Between moments there is nothing, no time, no space, just the infinite divisibility of another dimension.

GRAVITY IS INSTANTANEOUS

Therefore, as each new moment comes into existence, the world has to be reestablished. The universe must be re-formed. On a macro scale, we are talking of a reconstitution by the force of gravity. Thus, each moment, gravity has to be felt throughout the entirety; it has to travel the full extent of the universe, all 13.8 billion light-years (or 1.3×10^{26} meters), each Planck moment. Therefore, the true velocity of gravity would be:

$$1.3 \times 10^{26} \, m \, / \, 5.4 \times 10^{-44} \, s; \, or,$$
$$2.4 \times 10^{69} \, m/s.$$

This is equivalent to stating that gravity is instantaneous.

Now, this is what Newton alluded to; he felt that gravity, as a force, acted instantaneously throughout the entire cosmos. Einstein, however, described it as a curvature in space, and he felt that its actions, like that of all forces, would be constrained by the speed of light. Thus, to Einstein, gravity would be felt as if it were an electromagnetic wave—composed, not of photons, but of the very essence of space itself.

LIGHT'S VELOCITY—MEASURABLE BUT UNOBTAINABLE

The speed of light has always presented a significant conceptual problem. Nothing composed of matter can so move; it would become infinite in mass. Material things can only *approach* this velocity; they can never *reach* it. Also, at whatever speed something is moving, light would always pass it, in either direction, at the same constant speed. It became, to Einstein, the limit to our world, never to be reached by ponderous particles of matter, nor to be exceeded by massless messengers of force.

However, quantum theory allows, and innumerable experiments have proved, that there can be interactions (up to any conceivable distance) occurring instantaneously. So how can there be "immediate" entanglements, but, at the same time, a limit to their velocity? This is why Einstein felt that action at a distance—Newton's concept of gravity— was "spooky."

The solution lies in the third- and fourth-dimensional divide.

—

Our world is perceived as three-dimensional; however, it travels about a great fourth-dimensional curve toward the future. The totality of its three dimensions is reestablished by the universal force of gravity. It is the shaping force inside the equivalent of some gigantic particle—the universe. This same force traverses all quanta and exists within the very granules of space itself. Its velocity of resolution is 2.4×10^{69} meters/second. But electromagnetic waves travel from one moment to the next; they represent motion through the higher dimension. Since there are 8×10^{60} such moments (recordable since the Big Bang, or to 90^0 on a curve), their velocity is diminished to a measurable, but forever unobtainable rate:

$$2.4 \times 10^{69} \; m/s \; / \; 8 \times 10^{60}; \; or,$$
$$3 \times 10^{8} \; m/s.$$

Thus, there is really only instantaneous velocity; it reestablishes the entirety—all quanta—each and every Planck moment. But inasmuch as there are innumerable Planck moments, it is recorded at the speed of light. Therefore, because it is truly instantaneous, it is unobtainable; and no matter at what speed an object travels, it always passes that object at a constant, measurable, but unreachable rate.

So what, then, is quantum entanglement? It is the equivalent to gravity's immediate action. It is what occurs within the three dimensions of each Planck moment. However, it is not an electromagnetic wave; it is not a message, or an exchange of information from particle to particle.

LIGO—GRAVITATIONAL WAVES

The recent findings by the *Laser Interferometer Gravitational-Wave Observatory (LIGO)*, if substantiated, would prove that measurable gravity waves exist and, like electromagnetic waves, travel at c. Inasmuch as gravity is so much weaker than electromagnetism ($1/10^{39th}$ as strong), they can only be found by exquisitely sensitive devices, those capable of measuring a change in length of $1/1000^{th}$ the size of a proton (around 10^{-18} m). (In reality, the equipment is even more sensitive, since the change in length of the entire calculating system would, in fact, be 10^{-21} times that of the actual measuring rod.)

These findings, once corroborated, open an entirely new means of "observing" the universe. Not only are we now able to distinguish electromagnetic waves, we could also, using LIGO, measure a completely different type of cosmic entity; objects would become "visible" solely by dint of their gravitational force (those that, before, were entirely "black" now become discernible). However, if this finding pans out, what happens to the paradox of instantaneous versus measurable but unobtainable velocity? If gravity waves moving at the speed of light—c—exist, how is gravity able to exert its force instantaneously? If Einstein is right, is Newton wrong?

The answer is that both are correct. Gravity is instantaneous in its action—in its ability, from one Planck moment to the next, to reestablish the cosmos. If the entire universe is considered one gigantic particle, it, like all tangible entities, must disappear and reemerge each Planck moment. It cannot exist *between* moments as a three-dimensional object in a purely fourth-dimensional void. Therefore, gravity has to restore the world, instant to instant.

But we have seen that light, by moving at a measurable pace, travels through each moment to the next, particle to adjacent particle. The mathematics entails dividing the instantaneous velocity of a single moment by the innumerable moments involved; hence, both an immediate rate (over any distance), and a measurable but unobtainable speed, become one and the same.

—

RIPPLES IN THE FABRIC OF THE WORLD

LIGO was built to detect the minuscule change brought on by the gravitational wave's ripple effect in the structure of space. This exceedingly weak action is too feeble to instigate the movement of photons. However, I will argue in a later discussion that the whole concept of photons as massless carriers of force is incorrect, and that photons are actually three-dimensional particles—the remnants of larger entities ripped asunder by black hole encounters—flung deep into space to surround all atoms. Electromagnetic waves are equivalent undulations in the fabric of the universe, carrying energy from one source (a proton stripped of its neutralizing electron) to another. These wrinkles engage the "seas" of photons surrounding all atoms and cause the waves, which then accost their atomic shores.

If these ripples are exceedingly weak, they will not jar photons or lead to waves in an otherwise quiescent sea; we will not "see" or sense them. However, gravitational waves are not different from the much more potent ones easily discovered by conventional astronomy. They, and electromagnetic waves, are both composed of the "material" or granularity of the universe; nonetheless, because of their minimal intensity, they are ever-too slight to stir photons.

WITHIN AND BETWEEN

So what is really happening? Gravity is the reestablishment, Planck moment to Planck moment, of a single entity, the entire universe; it is instantaneous—2.4×10^{69} meters/second. Electromagnetic waves are the messages sent from one energy source (or unshielded proton) to others. These occur through instants of time, through the fourth dimension. Thus, their speed is measurable but unobtainable. It is an instantaneous velocity diluted by the great number of instants present. If we take the entire universe, the speed would be:

2.4 x 10^{69} m/s / 8 x 10^{60}; or,
3 x 10^8 m/s (the speed of light).

Therefore, if information (a decipherable message—an electromagnetic signal) were to be transmitted over an expanse of 3 x 10^8 meters, it would take 1 second. However, the instantaneous velocity (distance/Planck moment) over this span, if it were an immense, single particle, would be less than if the entire universe were covered; it would register at 5.5 x 10^{51} m/s (3 x 10^8 m / 5.4 x 10^{-44} s). But since, over this distance, there had been 1.85 x 10^{43} Planck moments (1 second / 5.4x10^{-44}s), or universes reconstituted, then the velocity becomes, once again, 3 x 10^8 m/s:

5.5 x 10^{51} m/s / 1.85 x 10^{43}; or
3 x 10^8 m/s.

Therefore, all individual particles, just like the totality, have an internal instantaneous velocity. It is the speed that allows for their realignment each Planck moment. Hence an electron, at 4 x 10^{-13} m (Compton wavelength), has a rate of about 7.4 x 10^{30} m/s (4 x 10^{-13} m / 5.4 x 10^{-44} s); and a proton, whose size is around 10^{-15} m, maintains one of approximately 1.8 x 10^{28} m/s (10^{-15} m / 5.4 x 10^{-44} s). Finally, if the message were to travel the slightest distance allowable, one Planck length (1.6 x 10^{-35} m), it would take one Planck moment (5.4 x 10^{-44} s), and its velocity, by definition, would be the speed of light—c (1.6 x 10^{-35} m / 5.4 x 10^{-44} s; or 3 x 10^8 m/s). (For a fuller discussion, please see Appendix H.)

Thus, the speed of an electromagnetic wave is constrained and constant; it is a result of the "bumping" of one Planck particle, or volume, by another, and then another, and so on, from the originating cause to the final effect. In many ways it is similar to a wave of sound (or pressure) moving through our atmosphere. The velocity within each particle is instantaneous; however, due to their minimal size (the smallest permissible), it is recorded at c.

The universe, then, because it is granular and moves with a centripetal motion through time, must depart and return each Planck moment. Force is propagated by direct contact—granule to granule—moving through each instantaneously; but, as there are so many, it becomes measurable yet never obtainable. Gravity waves and electromagnetism are the same: ripples in the fabric that makes up the three-dimensional world. Electromagnetism, being so much more potent (10^{39}, or one thousand trillion, trillion, trillion times), moves the bulky, orbiting photons that surround all atoms, and can be registered by our innate senses, sight and touch. Gravitational waves, as they are exceedingly weak, are now just being discovered by exquisitely calibrated devices (capable of determining changes as small as one part in 10^{21}). So both gravity (Newton's concept) and gravitational waves (Einstein's understanding) are the same; however, gravity is instantaneous *within*, and its waves measurable *between*, particles.

INTERCONNECTIONS

As a result, no matter what the distance covered, a message, particle to particle, one moment to the next, is moving at the constant speed of light, c (3×10^8 m/s). The quantum world is, on the other hand, entangled; it is found within the interstices of all particles—embedded in the instants of reestablishment, by gravity, of the three-dimensional world.

The universe is a giant 3-sphere; its individual but connected nodes are the innumerable (10^{80}) hydrogen atoms that comprise all. Therefore, since all nodes really represent the whole, all are reconstituted each Planck moment by gravity. Entanglement is the conceptualization of this intertwining of all hydrogen atoms, of all matter. However, to send a message, to get information from one source to another, to measure the electromagnetic characteristics of one entity by another, the force is externalized; it moves through "empty" space by direct contact—Planck volume to Planck volume—from moment to moment as the world recalibrates. It travels at c, the constant and measurable, but unobtainable, velocity of our world; and it and we forever age, as all must journey through the fourth dimension, through time.

Chapter 12

THE DEBRIS OF PHOTONS

Photons as particles of mass surrounding all

THOMAS YOUNG'S EXPERIMENT

TO RECAP: THUS FAR, WE HAVE EMPLOYED the fourth direction and the universe's granularity to explain some of the weirder aspects of quantum theory. However, the most important paradox of all is found in the double-slit experiment. This was first described by Thomas Young, an English scientist and true polymath (among his many accomplishments, he helped to decipher hieroglyphics) who showed, around 1800, that light was not, as conceived by Sir Isaac Newton, particulate (corpuscular) but instead consisted of waves. In this experiment, he split a light beam into two distinct rays that subsequently interfered with each other. In so doing, he proved beyond doubt that light, in a fashion similar to water, traveled with an undulating motion. This conclusion has since been confirmed innumerable times using light, electrons, and other entities, and the same inferences are always drawn: Something that may be transmitted as a particle will also interact as a wave. (We have, so far, maintained that wave–particle duality is the effect of higher-dimensional spin; I will soon attempt to show, however, that, in the double-slit experiment, the wave and particulate properties have a different basis.)

Before we begin, a quick journey through three-dimensional space is in order. We have described spin, a particle's centripetal action, as a motion through fourth-dimensional space, into and out of what exists. But particles truly reside in the real, three-dimensional world. There are definitely protons and electrons. All matter appears to have equal numbers of both—the universe is believed to be neutral. So what are we really describing when we speak of these particles?

ROTATION

A proton is a surface cover to a fourth-dimensional abyss. We have shown that it spins into and out of this higher dimension; in our world, however, it also has a distinct circular or rotating motion. Thus, in three dimensions, this component presents as rotation about an attracting, central core—an unfathomable black hole. We define it arbitrarily as a positive force, a pull toward that center. The electron is likewise rotating but, as it is situated at a greater distance from this core, does not need to move as fast to maintain its coherence. It presents a much more diffuse surface than does the proton, as it exerts much less resistance to capture. We arbitrarily give it a negative charge, exactly equal but opposite to the proton's positive value. If the proton represents the centripetal force of attraction, the electron is the centrifugal, or counterforce of repulsion. Every action must, according to Newton, have an equal and opposite reaction; thus, to the proton's inner pull, there is an electron's outer push.

Finally, way beyond the electron cover, we find the barely discernible neutrino. It acts as a distant outer edge to the proton–electron duo, the final outskirts of our hydrogen atom. What its true mass is, or even whether it has a slight charge, has not yet been fully determined. However, I believe that it will eventually be found to contain a very minimal negative charge, as it acts, essentially, as an extremely weak and evanescent electron. (In reality, there are three different neutrinos, but, in order to simplify, we will discuss them as if there were only one.)

So rotation, in three dimensions, is essential for existence; it counters the force of attraction inherent in all entities. But when an object rotates, its components must do so in only one direction; if they moved "helter-skelter," there would be collisions leading to annihilation. Rotation, besides being unidirectional, also causes a slight distortion in shape; areas at the midpoint move faster than those at the top or bottom. If we take the Earth as an example, something at the equator will travel 25,000 miles in 24 hours (over 1000 miles/hour), whereas at the poles it may travel less than one mile in the same period (the difference in velocity can be as much as 25,000 times). On our solid planet, this presents as a slight bulge at the equator and a minimal flattening at the poles. On the much more diffuse electron, we get a significantly widened midsection and a diminished overall height; it becomes shorter and fatter.

SURVIVING THE PLUNGE

Substances situated at the equator can easily resist the pull toward the core—their rotational speed allows for a successful centrifugal counterforce. However, as we travel toward the poles, this velocity constantly slows, and the centrifugal motion becomes ever-less able to resist the inward tug. Things, therefore, fall into this abyss; but some, just missing, can spew outward to either orbit (north–south) or escape into the great beyond:

—

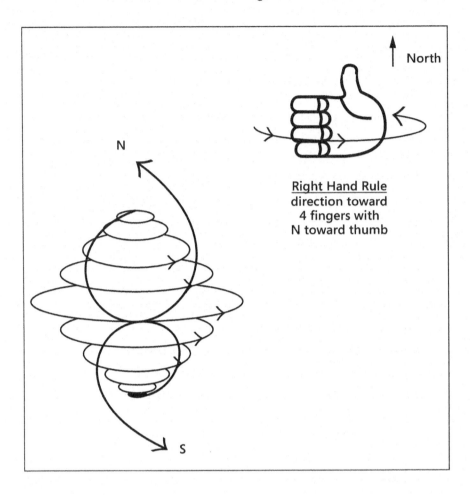

Hence we can visualize a rotation with, if we use a right-hand rule, east–west horizontal motion parallel to a surface (in the direction of the four fingers in the illustration), north upward (toward one's thumb), and south downward. These directions of movement—horizontal (approximately 30^0 +/- about the equator), and vertical (also around 30^0 +/- about the poles)—become the basis of electromagnetic waves, ripples in our world felt in the granularity of the smallest particles of space (Planck 2-spheres), spread outwardly from every center at light speed. However, between the equator's expanse and the polar regions there exists about 30^0 of surface (from more or less the 30th to the 60th parallels) where substance may or may not be stable. That which is only minimally so can be pulled inwardly and, in a fashion similar to what occurred at the poles, either engulfed or expelled to distantly orbit or completely disappear.

PHOTONS

Since the material pulled to the center is "electric" of some nature, we will assume it to consist simply of energy; therefore, when forcefully ejected (to be either entirely lost or orbit at a great distance), it would congeal into particulate matter. So the full picture of our atom contains a positive proton core pulling on a negative electron cloud with its distant, diffuse neutrino cover just barely enclosing the whole. The electron cloud is tugged inwardly but maintains its shape, and ripples from this cloud are felt throughout space as waves of granularity (through Planck 2-spheres). Finally, strewn over a great distance, we find material orbiting the entirety as a vast "sea" of dispersed, congealed energy—our photons.

I am therefore showing that photons, supposedly massless carriers of force, are in reality objects ejected from all atoms but retained as huge reservoirs of matter distantly orbiting their origins. They resemble the Kuiper belt and Oort cloud of frozen debris and comets encasing our solar system.

This view differs, I must point out, from that of most scientific approaches, which by definition assign electromagnetic force to entities without mass (photons) traveling as both waves and particles at the speed of light. However, I will demonstrate how, if adopted, my perspective helps answer the double-slit conundrum, and, further, how it explains the photon's spin (a complete turn takes a normal 360^0), the photon's lack of participation in the Pauli exclusion principle, and the apparent absence of an antiphoton; finally, it may even shed light on the riddle of dark matter. Let me, therefore, start, in the following chapter, with the strangeness of the double-slit experiment.

Chapter 13

DOUBLE-SLIT EXPERIMENT

Coming to terms with Thomas Young's famous experiment?

DOUBLE-SLIT WITH LIGHT

I N THE DOUBLE-SLIT EXPERIMENT, Thomas Young showed that light consisted of waves, not particles. However, just over a hundred years later, Einstein demonstrated, in his explanation of the photoelectric effect, that even if light was wavelike it was also composed of particles—now known as photons. So, today, we have both, a wave-particle "dilemma."

The double-slit experiment can be described as follows: A ray of sunlight shining through a small aperture in an exterior wall initially passes through two narrow slits of an interior surface and then appears on a second wall as an interference pattern. Since this experiment is only consistent with waves, it proved, conclusively, that light is so composed: Light is wavelike, not particulate. Let me illustrate this with the drawing on the following page:

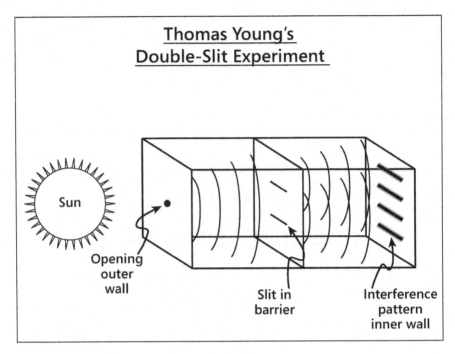

Thomas Young's Double-Slit Experiment

Sun

Opening outer wall

Slit in barrier

Interference pattern inner wall

DOUBLE-SLIT WITH ELECTRONS

Over the years this same experiment has been performed in-numerable times and in many different modalities. Probably the most significant trials were done with electrons. In these demon-strations, an electron gun shoots particles against an opaque, im-penetrable barrier with two openings, or slits, and their final impacts are recorded on a fluorescent screen. If one slit is closed, a simple clumping of electrons is found on the screen, similar to what individual particles would create; however, when both slits are open, one gets an interference picture. We therefore have con-flicting results showing *both* the wave and particulate nature of electrons.

When this experiment is then performed with both slits open, one electron at a time, we initially get single bursts on the fluores-cent screen as if only a solitary electron had struck. However, over time, these individual impacts coalesce into an interference pattern similar to what was found with light. So it makes no difference whether electrons pass the slits individually or *en masse*—they always eventually assume a wavelike nature.

Finally (and this is where the experiment gets stranger yet), if a recording device, essentially a light-emitting source of some kind, is placed right after the two slits (allowing the observer to discover by its "flash" which of the openings an electron had passed), and if only one electron is sent each time, those electrons remain particulate. They no longer form an interference pattern but, even over time, remain as two distinct clumps, just as if they were solid macro-particles (bullets, for example). Therefore, knowledge of which slit the entity uses maintains it as a particle; lack of the same allows for waves. And what is equally bizarre, if the information is not wholly accurate, if the recording device is not appropriately calibrated (that is, not fully sufficient in either frequency or intensity), the final result presents, concurrently, as both a clumping and an interference pattern.

The usual reason given for these weird findings is that an electron is both a particle and a wave, and can act as either, depending on the experiment performed; therefore, even when it could register as a particle, it can still present, if its actual trajectory is unknown, as a wave.

This has always been a most "unsatisfying" explanation.

SOLVING THE RIDDLE?

To solve these riddles, one needs to reexamine the entire basis of the just-described experiment. The electron gun employed contains a wire that has been heated in order, it is believed, to shoot out unstable or highly energetic electrons. I propose instead that, by heating this metal, the electrons are merely dislodged into the wire, not expelled from some gun; they are therefore no longer blocking their complementary protons (the core attractors); they no longer neutralize that inwardly directed force.

Since this newly freed force is really a pull toward the nuclear core (of the wire's now partially uncovered atoms), every time this lack of suitable protection (the hole where an electron had been) aligns more or less with the barrel, a force is felt (as if shot from a gun). Hence this force waxes and wanes as the now somewhat defunct electron cloud rotates; and, although considered a particle flying through space, it is in actuality only the tug felt toward the metallic core.

Thus, what is expelled really consists of a tightly coherent force of attraction—a high-frequency pulse—pulling toward the metallic wire's core and equivalent, in energy, to the displaced electron. Since it is a wave packet, it can easily transit as many slits as encountered. However, as it is so closely bound, due to the nature of the confining barrel, most of it simply passes through one slit; only a small percentage of each burst enters the secondary opening. Also, because of an inherent "jiggle" effect as this pulse traverses the barrel, the gun's aim is, inherently, somewhat erratic —it may fire at either slit or the wall itself.

Hence, it is not a physical particle that journeys from gun to screen to be recorded as a burst; it is a semi-coherent assemblage of force that impacts a sea whose waves lead to fluorescent changes.

If both slits are open and the gun is discharging numerous "electrons," the great number of collisions on this sea of photons leads to multiple waves—an interference pattern. If one slit is closed, only the single primary pulse of energy can pass, so only individual waves crash to the shore. This leaves clumping or a localized end product. If the gun is adjusted so as to expel but one electron each time, most of that individual grouping randomly enters one of the two open slits. However, that packet still has a residual or splatter effect that, even if only of minimal intensity, will traverse the other slit. This partial, subdued force causes only

—
90

a slight upheaval in the photon's sea, a disturbance not readily recordable as a fluorescent change. However, over time, both slits are arbitrarily entered and these minuscule ripples are energized into significant waves to form an interference pattern.

Finally, if a device is strategically placed between and just beyond the slits to record which one allowed the "electron" to pass, its flash, as the "electron" departs, is actually the interference of its light waves (at right angles) with the pulsed source of energy (the "electron"). It shows through which slit the major portion of that individual pulse traveled, as the other slit only encountered a small segment of this bundle (the mere residue uncoupled from the predominant, coherent, and aimed pulse). Nevertheless, that recording device's light source shines with equal magnitude on both slits, and its power is adequate to disrupt (yet not record) those meager secondary waves. Hence, only the major wave collection reaches the sea of photons (just a single wave is established per "electron"), leading to individual, non-interfering impacts—clumping—on the monitor's display. However, if this recording device is set too low (if its frequency or intensity is insufficient to "flash" the passing "electron"), it then, often, does not fully destroy the secondary waves; hence, interference patterns again emerge.

Therefore, although an electron is being recorded as passing an opening, in reality a tight, primary packet of wave energy is being registered. The documentation process, in turn, leads to dissolution of secondary bundles and allows for the equivalent effect of a slit closure. Hence, measurement, or knowledge, leads to "particles"; lack of the same yields "waves."

If one allows for the above interpretation, a coherent explanation of the double-slit experiment becomes feasible—its basis the movement of a force, or electromagnetic wave of attraction, that travels from the uncovered proton core of the metallic wire through the barrel of the gun, and through the slits, to impact that distant sea. The final change on the fluorescent screen is due to a tangible photon, of some mass, actually hitting a shore. It is a wavelike action, but a wave of material analogous to that of water.

FREQUENCY—PROPORTIONAL TO STRENGTH

Thomas Young's initial experiment was basically the same. The force (the shaft of light) moving from the Sun through the outer wall, which then went through two slits of an opaque inner barrier, was similar to our electron gun. The final impact causing interference was in the sea of photons. When Einstein, by unraveling the photoelectric effect, showed that light was particulate, he was describing this material sea (the greater its frequency of wave strikes, the more substantial the stirring of electrons).

There exists, in any electromagnetic or light signal, a minimal frequency (cycles per second) that is necessary to move distant stationary electrons—in order to trigger a photoelectric effect. The reason for this linkage (frequency and strength) is that the intensity of an electromagnetic wave is really a proxy for distance to core protons; the closer, the greater the attraction, and the faster the rotation. Hence, the more powerful the pull, the more "frequently" does a break or opening present and the higher would be the frequency.

Therefore, if one is willing to allow that, surrounding all atoms, there is a vast accumulation of fine, particulate matter—photons—an explanation of light's puzzling, composite nature can be given. However, it is not the wave–particle duality seen as entities spin between their real and virtual selves; it is a distinction of a different kind.

Chapter 14

DARK MATTER AND DUALITY

Shedding "light" (photons) on dark matter;
and duality's dual aspects

DARK MATTER

I F THE DOUBLE-SLIT EXPERIMENT makes sense with force, not distinct particles, then the photon becomes a particle of mass. It is a remnant of the electron cloud pulled toward the nuclear abyss that, upon just missing, is flung outward into a great, distant orbiting sea. We find something similar, as already mentioned, in the vast Oort cloud and the Kuiper belt of frozen debris—planetoids, asteroids, and comets—that surround our solar system.

If the analogy holds, a fine sea of particles can explain the riddle of *dark matter*. Since first mentioned in the 1930s, some kind of unknown mass or "dark" material has been postulated to accompany galaxies, allowing for their rapid rotation (a motion too fast to be accounted for simply by luminous matter). According to accepted theory, this material should be found in a "cusp" near a galaxy's center; however, this unseen mass appears to be present in a "halo" surrounding a galaxy's edge. This has been called the *cuspy-halo problem*.

The true etiology of dark matter is not known. All that *is* understood is that it has a great deal of mass (over five times that of visible objects); thus, it also has a huge gravitational effect. It does not have an electromagnetic signature (it does not emit appropriate waves—it is dark); could it, therefore, be composed of photons spewed out from the core of a galaxy to forever distantly orbit?

I am suggesting that a galaxy is similar in many respects to an atom. Its core is a great central black hole, surrounded by the equivalent of an electron cloud. We see this as the central bulge of many galaxies. The material spewed outward could, if coarsely grained, coalesce in spiral galaxies as the spokes of stars. Finally, the finest substance—the photons—could be flung much farther out to orbit as an unseen, vast sea of very minimal, particulate material: dark matter.

FERMIONS AND BOSONS

These suppositions are merely that; they are not proven reality. However, if one were to make the leap and accept the photon as a very fine particle of minimal mass (the missing remnant of the swirling energy pulled toward protons), then some other significant conclusions could be reached. As already noted, fermions (particles of matter) present with wave–particle duality, they obey the Pauli exclusion principle, they exhibit *zitterbewegung*, and they have spin 1/2; bosons (carriers of force) do not.

I am showing that a boson (in this case a photon), as a remnant of a fourth-dimensional spinning entity (an electron), is only a third-dimensional object, the condensate of energy, barely escaping the black-hole grasp, flung deep into space to coalesce into particulate matter. It is formed wholly and simply of this world; it is three-dimensional.

It, therefore, exists as you and I do, in recognizable three-dimensional space. It does not have a fourth-dimensional "mirror image" or antiparticle twin. It does not spin into and out of our plane. It does not partake in the wave–particle duality of real versus imaginary space. It has no spin in time (the fourth direction), so it does not exhibit the Pauli exclusion principle or *zitterbewegung*. Finally, since it is solely a "real" object, it rotates in a manner similar to that of any other physical entity—360^0.

THE "DUAL" NATURE OF WAVE–PARTICLE DUALITY

Furthermore, although the boson does not have the simultaneous physicality of a wave and a particle, it does form as a vast sea of material and, if breached by some force (the attraction, or tug exhibited by some electromagnetic impulse), it acts as would any equivalent fluid body—it forms a wave of particles. So the wave–particle duality, ascribed to a boson (photon), is due to a wave "through particles"; it is not due "to particles" that also somehow act as waves.

Although this conclusion may be quite a reach, I am not the only one with such unusual views. A very prescient article by the Chinese author Yang Shi-jia, in the science magazine *Infinite Energy* (January/February 2016), goes into the details of the difference between an electromagnetic wave (light) and the wave characteristics of a particle (in his example, a neutron). He shows that, when light traverses some medium (water, for instance), it slows, but upon exiting it resumes its journey at its normal speed—c. A rapidly moving neutron, however, also slows upon entering but, when leaving, does not go back to its original speed; the medium has permanently changed its velocity. Thus, the wave that presents to us as light is of a different character than that of a particle; or, as the author states, "the nature of light is [a] wave, and the character of [a] particle [or photon] is [the] derivative...

—

[whereas] the nature of a material particle [e.g., a neutron] is [a] particle, and the character of a wave is [the] derivative."

Thus, there is a dual solution to the wave–particle mystery. Electromagnetic waves are the forces transmitted from one Planck 2-sphere to another, through all of space, instantaneously but, due to the inordinate number of such spheres, at the forever unreachable, hence limiting, velocity of light. They interact with all material, specifically the vast, quiescent seas of photons surrounding each atom. These seas, in turn, batter the electron shores with continuous pulsations—with waves of photons.

The duality seen in an individual fermion (particle of matter), on the other hand, is the spin inherent in *all* such particles that allows for existence. It is the centripetal force of containment; it is what encases the unknowable fourth-dimensional abyss and what we perceive to be, at the same instant, both wavelike and particulate.

PART THREE

Final Formulations

Chapter 15

ROTATION AND THE REAL WORLD

*Rotation in 3D allows for
the counterforce of existence*

MAGNETISM

THE PREVIOUS CHAPTERS have been quite speculative, perhaps overly so. But if acceptable, they help to solve the riddle of dark matter (with its cuspy-halo problem) and explain the distinctions between fermions and bosons. Finally, in conjunction with the use solely of force, not actual particles, an explanation of the always difficult and mysterious double-slit experiment is possible.

If we now venture into a less contentious area, I want to provide a brief explanation that ties magnetism to a feature found in the arrangement of electrons—their stability in groups of eight. Magnetism can be seen as the polar aspect of the electron cloud that encircles an all-engulfing nuclear core. It can be drawn as a north–south flare combined with an east–west rotation, and is present at both poles:

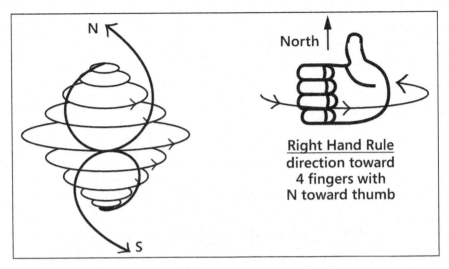

At either pole, there are both vertical and horizontal motions, whereas at the equator there is only a horizontal one. However, each pole presents slightly differently. In this illustration viewed from the equator's perspective, north shoots outward from the right and inward from the left, and south is the opposite—they both would be considered to be rotating counterclockwise. So if we draw a north–south attraction we get a natural fit. Both horizontal and vertical motions meld together:

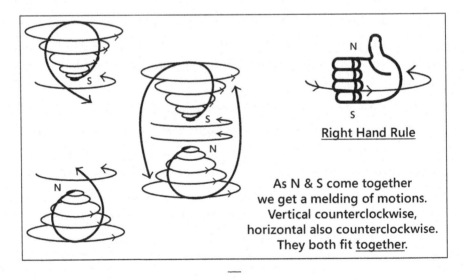

We now have a similar direction of horizontal motion (west to east), with a clean and neat interchange of vertical directions (counterclockwise). But if north approaches north, or south south, then we get the opposite: Both horizontal and vertical directions *impede* each other, and we have repulsion. So attraction or repulsion in a north–south orientation—magnetism—is secondary to rotation in both horizontal and vertical directions. When both mesh, they attract; when opposed, they repel:

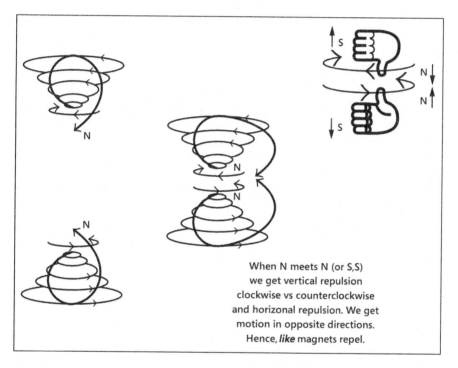

When N meets N (or S,S)
we get vertical repulsion
clockwise vs counterclockwise
and horizonal repulsion. We get
motion in opposite directions.
Hence, *like* magnets repel.

OCTET RULE

In a similar fashion, when entities come together in the horizontal plane, we get alignment if they move opposite to one another (clockwise and counterclockwise). Upon contact, both objects are travelling on the same route. This is the Pauli exclusion principle, but in three, not four, dimensions. So rotation in the "same" direction leads vertically to conjoining, whereas rotation in "opposite" directions permits horizontal contact. If we now want to build the most stable edifice, we can show that the *Octet Rule* (eight corners of a cube) allows for a true interlocking and maximum strength:

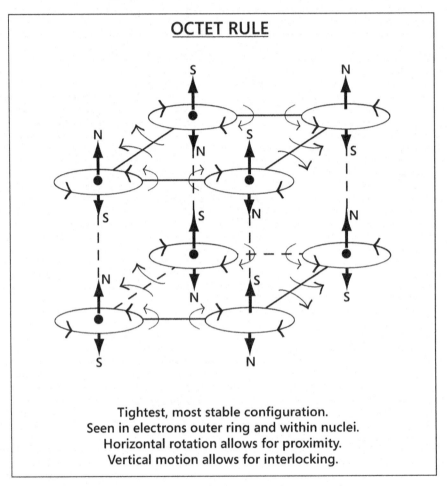

OCTET RULE

Tightest, most stable configuration.
Seen in electrons outer ring and within nuclei.
Horizontal rotation allows for proximity.
Vertical motion allows for interlocking.

PULL AND PUSH

Once we have the octet configuration, we can show why different atoms adopt it to join together. A similar occurrence is seen in the nuclear core, for, just as the electron's rotation allows for appropriate placement, the nucleons' rotations are similar. Both the proton and the neutron rotate; hence, both have magnetic force. In the nuclear core, a series of eight protons can also lead to the greatest stability.

Electromagnetic force, then, is dependent on the natural motion that all three-dimensional objects take under the stress of an ever-present central pull. This constant centripetal attraction becomes the universal force; its countering, by centrifugal push—inertia—is how the world is established.

Chapter 16

CONCLUSION

If not the Big Bang, then what; a summary and synthesis of the conundrum

DE-CONSTRUCTING

THIS BOOK'S IDEAS ARE AT BEST UNUSUAL. They certainly do not fit mainstream thought. But in their defense, they follow an internal logic. They are based on a picture of the universe as a great three-dimensional cover to an unknowable fourth-dimensional core.

If the universe is truly some immense 3-sphere, then Hubble's redshift paradigm does not connote expansion, it is only how measuring rods elongate with distance. The whole concept of a "beginning" needs to be abandoned—it is mere speculation; there is no reason to start with a Big Bang or any other initial event. The entire silliness of inflation consequently disappears; it is essentially impossible anyway. Dark energy, a new and unknown source of power needed to explain increasing expansion, is an illusion. Since there was never an expansion, distant celestial objects cannot be farther away than predicted; they are not pushed by some quintessential essence; instead, they are where they belong.

Ultimately, the Big Bang theory has become too unwieldy; it has too many patches, *ad hoc* additions, for sustainability. Even its most basic concept, that of the cosmic microwave background (CMB), can be more easily understood as the stretch in length one gets when measuring the smallest distance (Planck scale) at the greatest possible separation.

But once we discard the idea of a Big Bang, the Standard Model of particle formation also requires a serious restructuring. Although an elegant concept and one rooted in verifiable findings at lower energy levels, basing the current slew of particles on an initial cataclysm that expanded and slowly cooled is incorrect. The Standard Model never gives a reason for today's absence of antimatter. Baryogenesis, with its sister leptogenesis, cannot be substantiated; they are not founded in reality; they are mere artifices plucked from some fanciful ether, implanted out of necessity. As the rickety structure of the Big Bang collapses, the untenable construct of particle formation must also topple.

RE-CONSTRUCTING: GRANULAR SPACETIME

But if one does away with current beliefs, what can replace them? This book has, I hope, opened a vista to a dimension beyond what we commonly sense. To most, the next dimension is time, not distance. However, Einstein showed that space and time—space-time—are one and the same. We intuitively understand this when we mention being minutes or hours from a destination. We know that, when traveling at a constant speed, say forty miles an hour, our location may equally be thought of as an hour or forty miles away; they are equivalent. The universe, as it travels at the constant speed of light, can likewise be measured in space or time.

—

But we also know that the world is granular; Zeno and Planck showed this. It is composed of particles of spacetime. As it spins incessantly through time, it is constantly restructuring, instant to instant (nearly 20 million, trillion, trillion, trillion times per second). Each moment the entire world is reestablished. Since we consider the world to be a great particle, it, like all others, is held together by a centripetal attraction, or surface tension—by gravity. This force is felt throughout and acts instantaneously.

If we were merely two-dimensional, inhabiting a "never-ending" plane or 2-sphere wrapped about a constantly spinning giant three-dimensional orb, we would not visualize the third direction as distance, but as time. We would forever live in the present moment as our supposedly flat world continuously shifted, at an unvarying velocity, toward this other dimension. We would only regard the past as a distorted, redshifted version of the present; a lengthening caused by the curve of time. This view would cease in a far-distant past, a place we would mistakenly acknowledge as the "beginning."

If we were the smallest tangible objects of that surface, if we were protons, we would be the 1-spheres (circular structures) constituting its edge. If a higher, or three-dimensional being were to observe a slice through that 2-sphere, at any position in space or time, that individual would find a complete circle composed of minute 1-spheres or protons held together as if in a giant ring by the centripetal attraction of gravity. Finally, every proton in that ring would consider itself to be the very center of all, as its perception would end, in either direction, at 90^0 on what was, to it, an unfathomable curve.

In a similar manner, the real, three-dimensional world is composed of particles—protons with their electrons and neutrinos—combined as hydrogen atoms. These make up its surface. Each particle is a 2-sphere (enclosure of a globe) held together by its inherent surface tension or centripetal force. If, as just noted, a higher, or, now, fourth-dimensional being were to view a slice through our universe at any position in space or time, that individual would find, from its exalted perspective, a 2-sphere (the surface of some gigantic empty orb) composed of the smallest, stable, tangible objects—protons. (Just as before, any transection of a higher-dimensional sphere leads to a lower-dimensional one. However, unlike in the prior example, it is very difficult for us to see this, as we are part of the segment, and it would appear to come from a direction at a right angle to what we comprehend, from inside out. For a fuller discussion, turn to Appendix I.) As in the previous instance, this 2-sphere would be held together by a surface tension or force, by gravity. Finally, as in that scenario, each proton would consider itself the center of the universe, since the world would disappear in all directions at 90^0, the edge of an unknowable fourth-dimensional curve.

Thus gravity exists moment-to-moment, maintaining each surface universe (2-sphere) as it merges into the next. It establishes a giant enclosure that keeps all particles within its grasp. It is the overarching force that holds the great particle, our world, as one. Therefore, each individual particle (with its antiparticle), each piece of the hydrogen atom, although maintained intact, becomes part of that awesome structure.

RE-CONSTRUCTING: EXTERIOR–INTERIOR

Hydrogen atoms are the major components of our world. Helium atoms are far-and-away the next most common entity. Together, all other elements comprise the remaining one, or so, percent. However, the nucleus of each and every atom makes up the surface of our universe with gravity, as the attractive force between atoms, the glue that holds it all together. So, the entire complement of *fundamental forces*—the *strong force* within and between protons, *electromagnetism* grasping electrons to protons, the *weak interaction* involved with both the release of neutrons and the establishment of neutrinos, and *gravity* maintaining the entirety—are all aspects of the same centripetal attraction.

Since the surface of our 3-sphere universe is two-dimensional yet two-sided, as all surfaces must be, the "other" side represents the virtual aspect, "our" side the real. Each stable particle (proton, electron, and neutrino) has an "other-sided" virtual doppelgänger or antiparticle. The hydrogen atom, being the aggregate, is mirrored by the neutron, an inverse combination of the three. In order to maintain the surface integrity of the world, each particle constantly spins (centripetally) between the real and virtual. This continual spin is, at the same time, what gives rise to the quantum world's strangeness and the surface world's strength.

Finally, although the exterior of our universe is a two-dimensional facade (the surface of its particles), the interior of the universe is three-dimensional (empty space). But, of course, empty space is not "empty"; it is filled to every conceivable nook and crevice with the tiniest possible granular structures (Planck volumes, about 10^{180} in total). Each of these reestablishes itself, along with the universe, every instant. Yet as they are, by definition, the smallest size possible, their instantaneous re-formation occurs at the speed of light. Hence, vibratory waves, transmitted throughout the great volume of empty space, both electromagnetic and gravitational, travel with and through these particles as they collide, one against the other, and are recordable at the velocity of light—the limiting speed of our world.

SELF-CONSTRUCTING

Therefore, what I have been endeavoring to describe is a self-constructing schemata. *It exists because it exists.* The three-dimensional world is a granular structure. Each grain or Planck volume helps set its foundation. They form out of the necessity of existence. They are the minimal allowable contours to the fourth-dimensional abyss. The surface of each granule is composed of energy but measurable in its stability and coherence as mass (billions of electron-volts). The force that keeps each intact is transmitted to its neighbors, and they adhere, as if forming bubbles, to merge into distinct and tangible, larger objects—protons.

We measure the surface tension of protons as the strong force, but the same granules similarly self-construct into electrons (electromagnetism) and, finally, neutrinos (weak interaction). All these entities help comprise the surface of our 3-sphere, which, in its totality, is the cover to an unfathomable 4-ball. Thus, the material universe that we partake of, the atoms of which we are composed, is but a surface to a surface; it is the two-dimensional exterior of a 3-sphere enclosure. Empty space constitutes the immense interior or volume of our 3-sphere, greater than its facade by a magnitude of 10^{41} times. Finally, the four fundamental forces are merely manifestations of the surface tension of self-formed objects, given the inordinate, inherent attraction of each and every granule.

The universe, most simply put, is *made up of things that move through empty space.* It is replete with objects whose existence and motion demand granularity. These granules, or smallest surfaces (Planck volumes), in turn, inexorably coalesce and their surface tensions lead to the whole. The three-dimensional universe, therefore, because it *does* exist, is the *necessary cause of its own existence.*

FINAL CONSTRUCT: THE HIGHER DIMENSION

So what, then, have I been showing? I have pictured a higher dimension. At times, this vision is cloudy; a clear roadmap simply does not exist. But it is a truer guide than Lemaître's exploding atom. It explains the increasing redshift and CMB. It provides an alternative that allows for antimatter and the overall neutrality of the world. It builds a framework for a unification of forces based on the necessity for centripetal pull (and establishes the world as the centrifugal counter-push). It allows for instantaneous action, Einstein's spooky nemesis, and correlates it with the universal and limiting speed of light.

These concepts explain the inherently weird aspects of the quantum world. Are they final answers? Of course not. But they *are* rough paths that can be trod by much more astute minds to delve ever-deeper into the realm of a distant dimension. So let us rephrase the title of this volume: If you can explain it *in the real world,* you don't understand it (for, at the same time, it is *both real and imaginary*).

Appendices

T HIS SECTION CONTAINS SOME of the mathematics that I have left out of the book. Most of it is rather simple and should be easily understood by the average reader. It may help to explain certain topics.

APPENDIX A (Refer to p. 13)

A "sphere" is the surface; a "ball" is the interior. A surface is found using the formula for a 1-sphere ($2\pi r$) and multiplying it by an interior two dimensions smaller than what is enclosed. Thus, the simplest example, the formula for a two-dimensional circle (1-sphere) is $2\pi r$; but it is really $2\pi r$ multiplied by an interior two dimensions smaller, or, in this case, zero dimensions. Since a zero-dimensional object is a point (equal to 1), then ($2\pi r$) multiplied by (1) equals itself, or $2\pi r$. Similarly, in a fourth-dimensional object (4-ball covered by a 3-sphere) the formula for the 3-sphere enclosure would be $2\pi r$ multiplied by an interior two dimensions less than the 4-ball; that is, a two-dimensional interior, the area of a circle or πr^2. Thus ($2\pi r$) (πr^2) = $2\pi^2 r^3$, the surface of a fourth-dimensional object (3-sphere). A 3-sphere then is the product of the circumference of a circle and its area.

APPENDIX B (Refer to p. 22)

As noted in Appendix A, all surfaces, all spheres, are circles ($2\pi r$) multiplied by interiors two dimensions less than what is covered. Hence, the universe as a 3-sphere ($2\pi^2 r^3$) is really just a circle; and a tangent to it can be reduced to our simple example of a one-dimensional straight line.

Thus, if we take the universe of 13.8 billion light-years to be a quarter of a circle ($0^0 - 90^0$) and we divide it equally, we get constant distances two-dimensionally that continually elongate one-dimensionally. Let us, therefore, divide 1/4 of a circle into 5 equal portions; each is 18^0 ($90^0/5$) and equals 2.76 billion light-years (13.8/5). However, although these portions are always the same, the tangents increase as we go farther out:

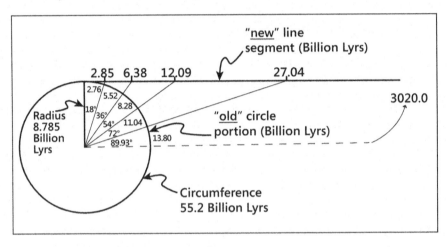

The tangent is defined as opposite/adjacent in a right triangle. "Opposite" is the new length on the straight line and "adjacent" is the radius of the circle comprising our universe. To find the radius all we need is the total circumference of our circle. Since 13.8 billion light-years is 1/4, the total is 13.8 x 4, or 55.2 billion light-years.

Now, $2\pi r$ = circumference, or (cross multiply),
r = circumference / 2π; since π = 3.14159,
2π = 6.28318;
therefore, radius, or r = circumference / 6.28318;

or, r = 55.2 / 6.28318; or, r = 8.785 billion light-years.

Thus, because, tangent = opposite/adjacent, tangent = straight line/radius, or (cross multiply),

tangent x radius = straight line.

Hence, all we do is multiply the tangent of each angle by 8.785 billion light-years:

angle	tangent	radius			line segment	circle portion
18^0	0.32492	x 8.785	=		2.85	2.76
36^0	0.72654	x 8.785	=		6.38	5.52
54^0	1.37638	x 8.785	=		12.09	8.28
72^0	3.07768	x 8.785	=		27.04	11.04
89.83^0	343.774	x 8.785	=		3020	almost 13.80

We now have the straight line segments and to get the z values we use the formula:

$$Z = new - old / old$$

where "new" is the line segment, and "old" is the circle portion (each 2.76 billion light-years). We now have a z for every tangent; thus, a z that matches each distance for the 4D curve/tangent column.

The distances for the Hubble/expansion column are based on the following formula:

$$Z = square\ root\ (1 + v/c\ /\ 1 - v/c) - 1.$$

To get the velocity (v) using Hubble's constant of expansion (current best estimate 67.8 km/sec for every 3.26 million light-years [megaparsec]) we take a distance (let us say 1.5 billion light-years) and divide it by 3.26 million to get the number of times its velocity has increased, then multiply by 67.8 k/s for each increase. Thus, in the example, 1.5 billion light-years is approximately 460 (1.5 billion/3.26 million) times 67.8 k/s, or about 30,000 k/s. We then put this velocity into the above formula:

$$z = sq\ rt\ (1 + 30,000 / 300,000 / 1 - 30,000 / 300,000) - 1;$$
$$thus,\ z = sq\ rt\ (1.1 / 0.9) - 1;\ or,\ z = sq\ rt\ (1.22) - 1;$$
$$or,\ z = 1.1 - 1;\ or,\ z = 0.1.$$

Thus, for any distance we can get a z value according to Hubble's law of increasing velocity with distance. When we compare the two columns we find distance to be greater for the 4D curve/tangent than for Hubble/expansion at every z until about 2.

Hence the universe is really bigger because of a fourth-dimensional curve than thought due to simple expansion. Dark energy, therefore, has no rationale. It is not required as there has been no initial or additional expansion. It is an illusion.

APPENDIX C (Refer to p. 25)

The edge at which the universe disappears—becomes invisible—is where a light wave would be so stretched as to be greater than the diameter of the universe, hence, no longer visible. Since the average light wave is about 550×10^{-9} meters (5.5×10^{-7} m), and the universe is about 1.3×10^{26} meters, if we divide these numbers we get:

$$1.3 \times 10^{26}\ m / 5.5 \times 10^{-7}\ m = .24 \times 10^{33},$$
$$or\ 2.4 \times 10^{32}.$$

Thus, once a light wave is stretched by a z factor of slightly more than 10^{32}, the wave is larger than the universe; it no longer is visible—the universe disappears.

APPENDIX D (Refer to p. 26)

The curve of the 13.8 billion light-years of our universe extends to a z factor of about 10^{32}. This represents the last 1.6 x 10^{-35} meters of that curve, the last Planck length possible. Given that all sites are equal each has an abundance of irregularities (anisotropies) consisting of galaxies and their clusters. Since a galaxy is about 100,000, or 10^5 light-years in size, and the universe extends just somewhat more than 10^{10} light-years, the difference is a factor of 100,000. Thus galaxies are 1/100,000th the size of the universe.

In a similar fashion, galaxy clusters range in size from 6 – 30 million light-years, or about 1/1000th the size of the universe. So the anisotropies are anywhere from 1/1000th to 1/100,000th of the cosmic microwave background (CMB) – which is more-or-less what has been found (1/10,000th the size of CMB).

APPENDIX E (Refer to p. 30)

The formula for centripetal force, the tension that holds the surface of a particle together is:

$$F = mv^2/r.$$
$$If\ v = c,\ then$$
$$F = mc^2/r,$$
$$and\ since\ mc^2 = E,\ or\ energy,\ then$$
$$F = E/r.$$

Now, energy represents the "essence" of the fourth dimension that must be contained. It exerts the same total intense "pressure" at all points (within the walls of all particles). Thus the smaller the radius of the particle the more concentrated is the force of enclosure. Gravity, the surface tension or force that maintains our world is based on a "particle" (the universe) with a radius of about 10^{26} meters, and a Planck volume has a radius of approximately 10^{-35} meters; therefore, the strength of the retaining wall or surface at a Planck volume is about 10^{61} times as great as gravity (as it is $1/10^{61st}$ its size).

Another way to visualize this is via mass. If mass actually is the density of the particle required to maintain its surface tension (to keep it spherical) the smaller the particle's size, the greater is its mass, as all particles must hold back the same total energetic force. Since a proton is approximately 10^{-15} meters, and the universe is about 10^{26} meters, or 10^{41} times as large, if the "surface" of the universe just consists of proton nuclei there should be more or less 10^{82} protons (10^{41} squared). The density of the entire universe then would be 10^{82} total protons in a volume 10^{78} cubic meters ($(10^{26}$ m$)^3$, or about 10,000 protons per cubic meter (10^{82} protons/10^{78} m^3). Since each cubic meter could hold 10^{45} protons ($(10^{15})^3$, yet contains only 10^4 it is diluted by a factor of $10^4/10^{45}$ or $1/10^{41}$.

Finally, if we take the product of the mass (recorded as billion electron volts) multiplied by the radius (meters), we get a "constant" of more-or-less 10^{-15} billion "electron-volt-meters," whatever that signifies, for all particles from the Planck volume to the universe itself, as is illustrated in the table on the next page.

Although this may just be an interesting exercise in numerology, it might also signify that the same total fourth-dimensional black hole "pressure" is exerted on all surfaces, and the smaller the enclosure the more robust or massive is the resistance.

	Mass	*Radius*
Planck volume =	10^{19} :	10^{-35}
Proton =	10^0 :	10^{-15}
Electron =	10^{-3} :	10^{-13}
Neutrino =	10^{-9} :	10^{-6}
Universe =	10^{-41} :	10^{26}.

APPENDIX F (Refer to p. 54)

The universe is a 3-sphere surface ($2\pi^2 r^3$) of a 4-ball. To us, in three dimensions, it appears spherical (equal, from any point, in all directions) and, therefore, its own surface presents as a 2-sphere ($4\pi r^2$). Thus there are 10^{80} (r^2) surface nodes or "actual" protons, but room for 10^{120} (r^3) "potential" protons in its volume. (We are using the radius of the universe as equal to 10^{40} protons for this exercise, not the more accurate number of 10^{41}; this loose usage is strictly for convenience.) Although the volume we are discussing appears to be empty space it is really composed of Planck granules and, thus, the virtual or potential protons are "lost" in this vast expanse.

Once, due to collisions, protons are deprived of their surface tension they no longer serve as an enclosure to the universe but dissolve into its interior, a space 10^{40} times as great as its surface (r^3 versus r^2). The universe can be viewed, therefore, as a gigantic particle; it has a surface or covering held together by a force or surface tension—gravity. This is the force felt between all particles (essentially, all intact protons). When a proton loses its coherence (when it and its antiproton meet or collide) it no longer is part of this surface and is "lost" in the great volume of "empty" space. Thus, it seems that for maintenance of the surface consistency of our world, a particle (proton) and its anti-self must, at all times, be secured on both sides of the world's exterior; when the particle and its twin meet on the same side that integrity disappears.

APPENDIX G (Refer to p. 68)

Zitterbewegung can be defined as the spin rate of an object or its cycles/second. If we take a simple example of a wheel with circumference of 5 meters and spin it around 2 times it will travel 10 meters. If we do that in 1 second it travels at 10 meters/second. Thus, if its velocity is 10 m/s, and its circumference is 5 meters, it has a spin rate of 2 cycles/second (velocity/circumference, or 10/5). An electron's velocity of spin is 3×10^8 meters/second and its circumference is 4×10^{-13} meters, thus, similarly to the wheel, its spin rate is:

$$3 \times 10^8 \; m/s \, / \, 4 \times 10^{-13} \, m \; = \; 7.5 \times 10^{20} \, cycles/second.$$

APPENDIX H (Refer to p. 78)

Instantaneous velocity is the speed attained traveling any distance in the least time allowable (one Planck instant, 5.4×10^{-44} seconds). Thus, whatever the distance, be it the entire universe (1.3×10^{26} meters) or one proton (10^{-15} meters), it has an instantaneous velocity.

Each Planck volume is 1.6×10^{-35} meters. Instantaneous velocity within each is, by definition, the speed of light or 3×10^8 meters/second; so instantaneous velocity is the same as the velocity of light if the distance is the least possible. The universe is made up of 8×10^{60} Planck volumes in any direction. Hence, each

—

Planck volume, or particle of space must move the adjoining one each 5.4×10^{-44} seconds, and the real velocity of *inter*-particulate matter is the speed of light, but of *intra*-particulate matter, instantaneous.

APPENDIX I (Refer to p. 107)

Any sphere, when fully transected, at any position, yields a sphere of a lower dimension; thus a 1-sphere (circle) when cut by a straight line leaves 2 points (0-sphere). Likewise, a 2-sphere (surface of a globe) when sliced by a plane becomes a circle (1-sphere):

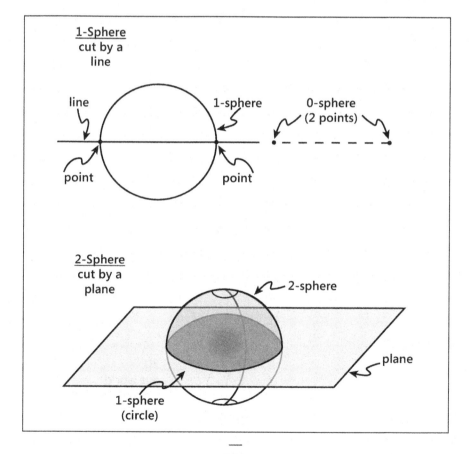

Finally, a 3-sphere (our universe) when intercepted by a three-dimensional "wedge" would yield a 2-sphere (a hollow orb). Since we are three-dimensional, if such a wedge of space were to cut though our 3-sphere world, we could not really observe the totality as we would be a part of it. It would appear to come from the fourth dimension, or from within all that existed. Viewing it, then, could only be done from that higher perspective.

Although we will make an attempt to visualize it (as a series of 2-spheres extending, one after another, toward the fourth direction), inasmuch as we really cannot envisage that higher dimension (as a direction) we will, instead, picture it as a series of 2-spheres moving forward in "time." Thus, each 2-sphere really becomes an individual "moment" of time. Let me make an attempt to illustrate this:

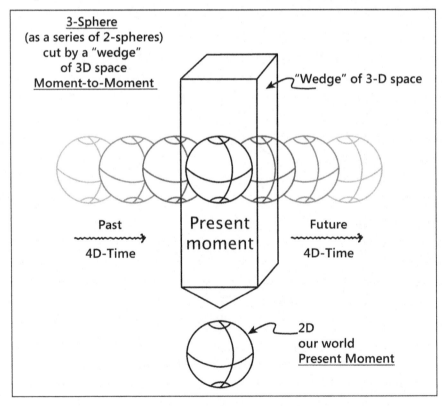